WALTER C. PATTERSON

NUCLEAR POWER

SECOND EDITION

PENGUIN BOOKS

Penguin Books Ltd, Harmondsworth, Middlesex, England
Penguin Books, 40 West 23rd Street, New York, New York 10010, USA
Penguin Books Australia Ltd, Ringwood, Victoria, Australia
Penguin Books Canada Ltd, 2801 John Street, Markham, Ontario, Canada L3R 1B4
Penguin Books (NZ) Ltd, 182–190 Wairau Road, Auckland 10, New Zealand

First published 1976
Reprinted 1976, 1977 (twice), 1978
Reprinted with Postscript 1980
Reprinted 1981
Second edition 1983

Made and printed in Great Britain by
Richard Clay (The Chaucer Press) Ltd, Bungay, Suffolk
Filmset in Monophoto Plantin by
Northumberland Press Ltd, Gateshead

To my parents,
who didn't worry when I dropped nuclear physics;

and to Cleone,
who didn't worry when I picked it up again.

Contents

List of Figures

Preface

Nuclear reactors are fascinating. They are the heart of the technology which may shape the world of our near future, or may obliterate it. Born amid the tightest of military security during the Second World War, nuclear reactors have always impressed the layman as esoteric, fantastic entities, beyond ordinary comprehension. Such an impression is unwarranted. As we balance on the threshold of total commitment to a nuclear pathway it is vitally important that nuclear policy be based on broad public understanding – of nuclear technology, of its applications, and of its implications.

Nuclear physics and nuclear engineering are, to be sure, specialized subjects, dealing in phenomena which sometimes seem almost Carrollian in their unexpectedness. But the essential features of nuclear reactors have changed very little in the four decades of their existence; what has changed is their size, and their context. The present book is an attempt to describe the reactors themselves – and to describe, too, the effect they have had, and are having, on the world in which we live. Of necessity the description is one man's view of an ever more controversial nexus of issues. It will strike some as unduly pessimistic about nuclear prospects, and others as entirely too absorbed in a subject from which they instinctively recoil. It is therefore appropriate to begin with a warning: in nuclear matters never take any one viewpoint as gospel, including this one. For those who wish to pursue questions further I have included in Appendix C (pp. 237–46) a lengthy list of additional sources, annotated to indicate – once again – one man's view of their virtues and shortcomings.

In the course of my own involvement with nuclear matters I have been fortunate to encounter many other viewpoints against which to measure my own, some in print and some in person. The staff of the United Kingdom Atomic Energy Authority have been unfailingly helpful to me in spite of our frequently diverging opinions. I should particularly like to thank Ron Truscott and his colleagues of the Public Relations section and Mrs Lorna Arnold of the Archives, who made available to me advance copies of Professor Margaret Gowing's superb official history *Independence and Deterrence* at a crucial stage in my own research.

The now sub-divided United States Atomic Energy Commission has supplied me with many useful documents, as have the International Atomic Energy Agency and the OECD Nuclear Energy Agency, the last through the good offices of Bruce Adkins, to whom again my thanks. I have made extensive use of the works of Dr Samuel Glasstone, the late Dr Theos Thompson and Dr J. G. Beckerley, the late Dr Kenneth Jay, my friend and colleague Sheldon Novick, Dr J. E. Coggle, Dr Tom Cochran, Dr John Gofman, Dr Arthur Tamplin, Dr John Holdren, Richard Lewis, John McPhee, Norman Moss, Dr Peter Metzger, Roger Rapoport, Ted Taylor and Mason Willrich, among many; to all of them my thanks. The publications of the Stockholm International Peace Research Institute, the Pugwash Conferences, the International Commission on Radiological Protection, the National Radiological Protection Board and the Union of Concerned Scientists, and the pages of the *Bulletin of the Atomic Scientists, Science, Environment, Energy Policy* and *Nuclear Engineering International* have yielded much valuable material, as has the Weekly Energy Report, to whose editor and publisher Llewellyn King my thanks.

Many of my journalist colleagues have for many months carried on with me an unending and fruitful colloquy, not only about nuclear affairs but about their context in energy and social policy overall. I expect they will let me know what they think of my effort, in similarly blunt language: in anticipation my thanks to the lot of them, especially to the staff of the *New Scientist*, who encounter my opinions sooner and more often than most.

Even closer to home my colleagues in Friends of the Earth International, without whose active participation this book neither would nor could have been written. Brice Lalonde and Pierre Samuel of Les Amis de la Terre in France, Lennart Daleus of Jordens Vänner in Sweden, Brian Hurley of FOE Ireland, Holger Strohm of Die Freunde der Erde in West Germany, Kitty Pegels of Vereniging Milieudefensie in the Netherlands, Jim Harding of FOE Inc. in the USA, and many other Friends have provided an international link-up of growing strength. In Britain Dr Peter Chapman of the Open University and Gerald Leach of the International Institute for Environment and Development have contributed provocative insights and stimulating debate. My Friends of the FOE Ltd office in London have put up with my absence from the team for some weeks without complaint, carrying for me my share of the load. Finally, and most particularly, my warmest thanks to Dr John Price and to Amory Lovins, with whom it is an exhausting pleasure to work, and without whom I might as well stick to cultivating my broccoli.

I am grateful to Peter Wright of Penguin Books for the opportunity to amplify fourfold an earlier version of this text. On the home front my thanks and best wishes to Mrs Sue Hunter, who has toiled nobly to turn a tangled typescript into intelligible copy. Lastly to my beloved wife Cleone, who put up with three months of me lying obsessively awake at four in the morning: I promise that I shall never again write about terrorists with nuclear weapons the day before Christmas Eve. I hope.

Walt Patterson
31 December 1974

Preface to the Second Edition

The most difficult task in preparing this second edition has been to decide what material from the first edition to leave out, to make the required room for new. The choice of material to include and to exclude is of course the responsibility of the author, underlining the comment in the first edition: 'Never take any one viewpoint as gospel, including this one.'

I am grateful to Penguin Books and Rab McWilliam for the opportunity to revisit at length this endlessly fascinating if often unsettling territory. My thanks also to Gerald Leach, who taught me how not to be overwhelmed by my subject matter, and to Liz Payne, Hilary Roy and Rita Morgan, who have kept me from being buried by it.

Walt Patterson
September 1982

Introduction:
The Nuclear Predicament

For four decades the world has been learning to live with nuclear energy. The learning process has been exciting, frustrating, and sometimes frightening; it is far from over. Indeed it may be just beginning. We have learned a great deal about how to release nuclear energy; how to control it; and how to make use of it. We have even learned to take it for granted. But we have not yet learned to live with it. Nuclear energy in all its aspects is already shaping the world. The future of our globe will depend to a startling extent on what we know about nuclear energy, and what we do about it. The crucial decisions will not wait another four decades.

Concentrated high-quality energy has become a staple commodity in our industrial society. The most concentrated energy available is nuclear energy, made accessible by nuclear reactors. The energy contained in 1 kilogram of uranium, if it were all to be released in a nuclear reactor, would be equivalent to that produced by burning some 3000 tonnes of coal. It is not, of course, quite that simple; the possibly apocryphal British workman who filched some uranium reactor fuel and tried to burn it in his grate was disappointed. But there is no doubt that the world's reserves of uranium represent a staggering store of energy. If suitable reactors and other facilities are provided, it becomes possible to exploit uranium, which would otherwise be virtually useless. The same is true of the even more plentiful metal thorium.

These constructive possibilities were identified very early in the development of nuclear energy, even as plans were taking shape to release nuclear energy explosively. The awesome destructive power of nuclear weapons dominated the scene for the first post-war decade. But by the mid 1950s scientists and engineers were well on their way to harnessing this power for peaceful purposes. The prospects looked brighter and brighter. There had, to be sure, been a surge of euphoric predictions in the aftermath of the two nuclear explosions over Japan which ended the Second World War. So-called 'atomic power' would run a car on an engine the size of a fist; we would soon live in houses heated by uranium; 'atom-powered' aircraft would be able to remain aloft indefinitely; 'atom-powered' rockets would enable us to cross the ocean in three minutes – and so on. But the people who really understood the implications of nuclear energy were much more realistic. They chose applications whose development seemed fairly straightforward; and their efforts bore fruit.

A nuclear reactor releases nuclear energy in the form of heat; the heat is used to generate steam, and the steam to generate electricity – with conventional electrical equipment. Since the mid 1950s nuclear generation of electricity has become a full-fledged technology. From the outset it seemed that the nuclear approach to electricity generation would have certain advantages and disadvantages, in comparison with conventional generating stations which raise steam by burning coal, oil or gas. Fossil-fuelled power stations are less expensive to build than comparable nuclear power stations. On the other hand, it was expected that the running cost of a nuclear power station would be considerably less than that of running a fossil-fuelled station. Some early publicity went so far as to declare that nuclear electricity would be too cheap to meter. But as usual those on the inside made no such claim. Instead they calculated the total cost of a unit of electricity generated by a fossil-fuelled or by a nuclear station, taking into account both capital and running costs. Estimates differed slightly, but there was every likelihood that a unit of nuclear electricity would cost only about one-fifth as much as a unit of fossil-fuel electricity. On this basis nuclear power stations looked an excellent investment.

In the ensuing years, the bases for these economic calculations varied. For a time the cost of oil remained low, while that of coal increased; some nuclear costs also increased, and the balance remained uncertain. By the late 1960s mounting public concern for the environment was drawing attention to the problems arising from large-scale use of fossil fuel: health hazards from underground coal-mining, ecological damage from surface mining, marine pollution from transport of oil, and air pollution from the burning of coal and oil. By contrast nuclear power stations seemed environmentally inoffensive.

In the early 1970s the upward surge of oil prices, and increasingly uneasy labour relations in the coal-fields, added to the comparative economic attractions of nuclear power. The gradual and tentative industrial commitment to nuclear power for a time accelerated dramatically; so did the nuclear component of total electricity output.

Governments wanted to lessen their dependence on the petroleum-exporting countries; electricity supply systems wanted to lessen their dependence on coal, especially because of their vulnerability to recalcitrant unions. Nuclear electricity generation seemed the obvious alternative. In the longer term, it was argued, coal and oil would become irreplaceable raw materials for the chemical industry, and should be reserved for these uses, while nuclear energy was used for electricity. Electricity, it was further argued, is a premium form of energy, versatile, high-grade and clean at the point of use. It ought accordingly to make up an ever-larger proportion of the total energy used. Since nuclear sources could most readily be used to provide electricity, it all seemed to fit together very neatly. World energy use would continue to rise rapidly; so would energy use per person, as more and more people shared in the benefits of modern technology. One authoritative view foresaw a world in which world-wide energy consumption per person would be twice that of present-day Americans – this energy would be provided by some 4000 clusters of nuclear power stations, each cluster containing enough reactors to produce five times the output of today's largest power station. For such a high-energy future the role of nuclear energy would be crucial. Only by the most vigorous possible growth of nuclear capability could humanity's energy requirements be met.

Such an argument was in many ways persuasive. It was not, however, unanswerable; and while some voices were calling for more and larger reactors as fast as possible, other voices were asking other questions, some of which still remain difficult to answer.

The earliest questioning derived from lingering public fear and distrust of nuclear energy, because of its first appearance as the most devastating weapon ever used. Gradually, certain specific issues crystallized out of the general unease. The world has somehow got used – albeit with deep unease and recurrent protest – to the overwhelming destructive power stored in the nuclear arsenals of the USA, the Soviet Union, the UK, France and China; few would hesitate to identify these arsenals as the most terrible threat to the future of life on earth. But apart from these explicit military aspects of nuclear energy, several other aspects also give rise to concern. We shall examine these in some detail in the coming chapters. It must here suffice to mention them briefly, as issues which will recur repeatedly in later discussion.

Nuclear reactors and other nuclear facilities produce and contain enormous quantities of material which is 'radioactive' (pp. 27–8). Some radioactive materials are very dangerous to living things, and may remain so for unimaginably long times. These materials must on no account be allowed to escape in quantity from nuclear facilities. Such facilities release small amounts of radioactivity to their surroundings during normal operations. One area of bitter controversy concerns the standards and controls applied to these releases. Some critics with impressive credentials consider present standards far too lax, especially if there is to be a continuing increase in the number and size of nuclear installations. Another major concern is operating safety, not only of the various designs of reactor themselves but also of their support facilities, including transport systems. It has become unpleasantly clear that such safety must take into account the possibility not only of accidents but also of sabotage, and even military attack. A protracted, expert disagreement about safety has plagued the most popular design of reactor, and continues unresolved. Other designs have not thus far been subjected to such intense independent scrutiny.

One category of radioactive material arising from nuclear activities requires particular attention. This is the 'high-level' waste from used reactor fuel (see pp. 94–100). High-level waste contains large amounts of substances which are dangerously radioactive, and will remain so for hundreds of years. What to do with these wastes is a question as yet unanswered. Provisional answers have been proposed, and interim management is said to be adequate, but in the long term the question becomes one not of technology but of ethics. Should we create these dangerous substances in ever-increasing quantities, to leave them to our remote descendants?

Ethics aside, it has become abundantly apparent that considerations of safety affect the overall cost of nuclear power. Just as coal-mining must take account of the cost of health measures, land reclamation and pollution control, the use of nuclear power must allow for costs of extra safety measures and related provisions. The optimistic cost comparisons originating in the early 1950s, which favoured nuclear over fossil fuels, no longer look unarguable, as we shall describe.

Expenditure alone, however massive, seems unlikely to provide the requisite guarantees for one aspect which might narrowly be related to safety. As the world economy comes to rely more and more on nuclear reactors as a source of power, so traffic in 'fissile' materials increases – materials which can be made into nuclear weapons. Authoritative studies have shown that provision for the security of these materials is all too often perfunctory. The prospect of nuclear weapons in the hands of unstable governments, terrorist organizations or deranged fanatics is not one calculated to encourage a rosy view of the global future. Dealing with this danger may entail special government nuclear police forces, relentless official probing of personal histories of nuclear employees, monolithic central administration of public policy, and other activities which come uncomfortably close to outlining a virtually totalitarian social structure.

It becomes apparent, when considering these intractable problems, that the decisions we make about nuclear energy will determine in large measure the kind of world our grandchildren will inherit. The issues this technology has created constitute a remarkable microcosm of the present predicament of our planet. The nuclear predicament raises a host of social, political and even ethical problems, many of them with long-term implications beyond any foreseeable horizon. Clearly such issues demand the fullest public consideration, the widest possible participation in the crucial decisions to come.

Public participation in nuclear decision-making has for too long been either tentative or desperate, largely because the issues have seemed to be cloaked in the most esoteric scientific obscurity. But the veil of mystery surrounding nuclear matters has always been primarily one of military secrecy, not of intellectual inaccessibility. In the next three chapters we shall describe why and how nuclear reactors work, and the other services they require. If you are a nuclear engineer you can skip these chapters. If not, you should read them carefully, as they make it easier for you to determine whether nuclear engineers are talking sense.

PART ONE

The World of Nuclear Fission

1 · What is a Reactor?

Atom and Nucleus

If you take a pair of metal hemispheres and slam them together very fast face to face, one of two things may happen. You may get a loud clunk. Or you, the hemispheres and everything else in the vicinity may be almost instantly vaporized in a burst of incredible heat. If the latter happens, you can be sure that the metal was a particular kind of uranium, not that the confirmation will do you much good.

What has vaporized you is raw energy, released from the innermost structure of the uranium. The energy in the interior of uranium was revealed to the world on 6 August 1945, in the sky above Hiroshima, Japan. Never has a source of energy made a more horrifying debut. Yet, paradoxically, the most overpowering energy humanity has learned to release comes from the tiniest reservoir we have yet learned to tap: the nucleus of an atom.

What is an 'atom'? And what is its 'nucleus'? Suppose you take a lump of lead, and cut it into smaller and smaller pieces. When the pieces are so small that your knife is too clumsy, switch to an imaginary knife and keep cutting. Ultimately the pieces will get so small that if you cut any more you will not get two pieces of lead: the next cut will change the identity of what you are cutting. The smallest piece which is still lead is called an atom of lead.

The word 'atom' means 'indivisible'. You cannot divide an atom of lead and still get lead. But you can divide the atom and get smaller pieces which are no longer lead. If you start to dismantle an atom, the first part you remove is called an 'electron'. Until this stage you have been able to cut without encountering any conspicuous electrical effects; but the electron is negatively charged, and the remainder of the atom is left positively charged. The parts of the atom have become 'ions': a negative ion (the electron) and a positive ion (the remainder of the atom). Each time you remove another electron you leave still more positive charge on the remainder. It becomes doubly 'ionized', triply 'ionized', and so on. Since negative and positive charges attract each other, it is harder and harder to prise off successive electrons.

Suppose, however, that you manage to remove *all* the electrons. (For most atoms this is in practice very difficult.) What you have left is the innermost heart of the atom: the nucleus. This is where all the positive charges are. Furthermore, you would now find an enormously increased difficulty in cutting any more. Surprisingly enough, although the nucleus contains only positive charges (which *repel* each other), its constituent pieces cling together with a loyalty which makes the outer electrons look frankly promiscuous.

There used to be much popular talk of 'splitting the atom', but the problem was rather one of splitting the *nucleus* of the atom. 'Atomic' bombs should have been called 'nuclear' bombs: for the shattering energy they released came from the rupturing of nuclei (one nucleus, two or more nuclei).

Uranium

What makes uranium so dramatically different from other substances? To appreciate its unique characteristics we must first consider some basic nuclear physics: that is, what nuclei consist of and how they behave. An atom is made of electrons around a nucleus; a nucleus in turn is made up of 'protons' and 'neutrons'. A proton has a positive charge; a neutron has no electrical charge and is 'neutral'. At first it seems difficult to understand how a nucleus stays together at all. The positive charges of the protons ought to push them violently apart. But within the compact volume of the nucleus a new kind of force comes into effect: an immensely powerful short-range attractive force acting equally between protons and neutrons – which, from this point of view, are all 'nucleons'. The short-range nuclear force holds them together, against the repulsive effect of the protons' positive charges. In this way the neutrons act as 'nuclear cement'.

However, in a nucleus which contains 92 protons – that is, a nucleus of uranium – the repulsive force among the protons is on the verge of overcoming the nuclear force. If there are as many as 146 neutrons also present, the nucleus can remain intact – barely. This form of uranium, containing in all 238 nucleons, is called uranium-238 or $^{238}_{92}U$. For reasons that need not concern us here, involving the grouping and compatibility of nucleons, the next most probable arrangement is a uranium nucleus containing three fewer neutrons: uranium-235, $^{235}_{92}U$. Atoms with these lighter nuclei make up about 0·7 per cent of naturally-occurring uranium. (If nuclei have the same number of protons, they are nuclei of the same chemical 'element': thus, every nucleus with 92 protons is the nucleus of an atom of uranium. Atoms whose nuclei have the same number of protons but different numbers of neutrons are called 'isotopes' of the element: for instance, uranium-238 and uranium-235 are isotopes of uranium.) The uranium-235 nucleus has a property unique among all the more than 200 types of nuclei found in nature in significant quantity before 1942. The uranium-235 nucleus is already under near-disruptive internal stress; a stray neutron blundering into it can rupture it completely.

Radioactivity Produces Radiation

When a stray neutron hits a uranium-235 nucleus, the result is a 'compound nucleus' of uranium-236. It is called a compound nucleus because it does not last long. The energy added by the neutron – even a 'slow' one – overcomes the precarious stability of the nucleus and, almost instantly, it flies apart. The rupture of a uranium-236 compound nucleus usually results in about two fifths of the nucleus flying off in one direction and about three fifths in the opposite direction, with perhaps two or three odd neutrons also shooting out. The flying fragments burst out with so much energy that a subsequent tally of masses reveals a shortage: some of the mass of the original nucleus has been converted into energy. This is the source of the enormous energies released in such nuclear events.

For example, one common subdivision results in one chunk of 38 protons and 52 neutrons, another of 54 protons and 89 neutrons, and 3 odd neutrons: making up, of course, 236 nucleons in all. The chunk containing 38 protons is a nucleus of strontium; since it contains in all 90 nucleons it is the notorious strontium-90. The chunk containing 54 protons is a nucleus of the inert gas xenon; since it contains in all 143 nucleons it is xenon-143.

Such a complete rupture of a nucleus is called a 'fission', by analogy with the biological term for the division of a growing cell. More precisely it is called 'nuclear fission'. When it is provoked by the impact of an additional neutron it is called 'induced fission'; such is the case with uranium-235 just described. Some very heavy nuclei are so unstable that they may rupture even without being struck by a neutron; such a rupture is called 'spontaneous fission'. Fission, whether induced or spontaneous, is the most violent kind of breakdown that a nucleus can experience. But there are others. A nucleus of uranium-238, for instance, while not being so near to rupture as its lighter relative, is still under severe stress: so much so that sooner or later it is likely to squirt out a lump made up of two protons and two neutrons. Since this makes a larger proportional reduction in the proton contingent than in the neutron contingent, the remaining nucleus, now containing only 90 protons and 144 neutrons, is slightly less stressed. (It is a nucleus of the metal thorium-234.) The lump or 'particle' squirted out is identical in every respect to an ordinary helium nucleus; but since it emerges with considerable velocity, and ploughs a furrow through whatever slows it down, it is given a special name: it is an 'alpha particle'. Most nuclei with at least 83 protons undergo a breakdown this violent; they are called 'alpha-emitters'.

The balance between protons and neutrons in thorium-234, while more satisfactory, is far from ideal. In effect, by emitting an alpha particle, the nucleus has overadjusted. This leads to a yet more delicate form of breakdown. Out of the nucleus containing 90 protons and 144 neutrons there suddenly squirts an electron. It is identical in every respect to the electrons outside the nucleus; but since it emerges with considerable velocity it too is given a special name: it is a 'beta particle'. The nucleus which remains now contains one more positive charge than it had. But since an electron is very much less massive than a nucleon, there are the same number of nucleons as before: 234. A neutron has apparently turned into a proton. The nucleus now contains 91 protons and 143 neutrons; it is a nucleus of protactinium-234. Like thorium-234, protactinium-234 is also a 'beta-emitter'; when it emits a beta particle it becomes uranium-234, which is an alpha emitter. So, leap-frogging down by alternate alpha and beta emission, the nucleus alters itself until it has only 82 protons and 124 neutrons, and is at last stable: a nucleus of lead-206.

On the way, the nucleus regularly finds itself, after emitting an alpha or beta particle, still unduly agitated or 'excited'. To settle itself down it gives off a burst of energy in a form closely akin to ordinary light, but much more energetic, and invisible. This burst of energy is called a 'gamma ray'. It is identical in every respect to the well-known 'X-ray', except that an X-ray comes from the electron layers outside the nucleus, whereas a gamma ray comes from inside the nucleus.

Consider also the nucleus of strontium-90, one of the two large fragments formed by the induced fission of uranium-235 in the earlier example. The strontium-90 nucleus, a 'fission product', has a disproportionately large number of neutrons for its protons, coming as it does from a much heavier nucleus which requires more 'cement'. Accordingly, the strontium-90 nucleus is also a beta emitter. Sooner or later it squirts out a high-velocity electron – a beta particle – and one of its neutrons is replaced by a proton. It becomes a nucleus of yttrium-90, another beta emitter, which by the same process becomes a nucleus of zirconium-90, which is stable. Beta emissions from fission-product nuclei are often followed by one or more gamma rays.

There are thus four ways in which a nucleus can alter itself: fission, alpha emission, beta emission, and gamma emission. From a lump of material containing such unstable nuclei the emissions from these activities shoot out radially in all directions: the lump is said to be 'radioactive', and the emissions – neutrons, alpha and beta particles, and gamma rays – are called 'radiation'. A collection of nuclei which shoots out one such emission per second is said to exhibit one 'becquerel' (Bq) of radioactivity, after Henri Becquerel, who first discovered the phenomenon of radioactivity in 1896. The becquerel is now the accepted international unit of radioactivity; but you will still see an older unit, the 'curie'. A curie of radioactive material shoots out not one but 37 000 000 000 emissions per second. This is the radioactivity of one gram of radium, one of the first substances known to be radioactive, which was discovered by Marie Curie. A curie is a lot of radioactivity; you will also encounter metric subdivisions – the millicurie, microcurie, nanocurie and picocurie, in descending steps of 1000 as usual. You can see that one nanocurie equals 37 becquerels of radioactivity. You should note also that 'radioactivity' produces 'radiation'; the two terms are not interchangeable, although even official statements sometimes use 'radioactivity' when they mean 'radiation' and vice versa.

In a radioactive substance it is impossible to tell whether a particular nucleus is on the point of radioactive breakdown, or 'decay'. Nonetheless, in a sufficiently large sample of any particular radioactive nuclear species, or 'radioisotope', a certain fraction of the nuclei always decay in a quite regular length of time. For instance, if you start with 1000 nuclei of strontium-90, 28 years later 500 will have decayed and you will have 500 left. After a further 28 years, 250 of the remaining 500 will have decayed, and you will have 250 left. And so on: however much you start with, 28 years later half will have decayed and only half will be left. Obviously the corresponding radioactivity will also have fallen by one half. For strontium-90 the period of 28 years is called its 'half-life'. Each radioisotope has a half-life for each form of radioactivity it exhibits: in each case the half-life is the time during which half the nuclei in a sample decay, and the corresponding radioactivity falls to half the initial level. Half-lives of different radioisotopes range from fractions of millionths of a second to millions of years.

The Effects of Radiation

Unless radioactive decay takes place in a vacuum, the radiation emitted must pass through the surrounding substance. The consequences depend on the substance, on the type of radiation, on its energy and on its intensity. An alpha particle, made up of four nucleons with two positive charges, interacts vigorously with surrounding atoms, tearing off electrons and knocking nuclei out of place. In doing so, the alpha particle quickly gives up its own energy, travelling only a short distance but doing enormous damage along its path. Most alpha radiation is stopped within the thickness of a single sheet of paper. A beta particle, much less massive and with only one negative charge, disturbs and dislodges neighbouring electrons, but loses its energy less swiftly and therefore travels somewhat farther than an alpha particle. Most beta radiation is stopped within the thickness of a thin sheet of metal. A gamma ray, with no electrical charge, loses its energy much more gradually, and can travel a long distance, causing a relatively small amount of disturbance at any particular point on its path. A neutron, also without electrical charge, is likewise free to travel a long distance, and is slowed down mainly by direct collision with nuclei. Gamma or neutron radiation can penetrate more than a metre of concrete.

Dislodging an electron from an atom makes the atom an ion: so emissions from nuclei are 'ionizing radiation'. When ionizing radiation passes through a material it causes changes in the structure of the material − sometimes temporary, sometimes permanent, sometimes useful, sometimes harmful. The effects of ionizing radiation depend roughly on how much energy the radiation releases into a given amount of material – the more energy, the more disruption. The original unit of radiation exposure was the 'roentgen', named after Wilhelm Roentgen, discoverer of X-rays.

The effects of ionizing radiation become particularly important if the radiation is passing through living matter; the delicate molecular arrangements of living matter can be easily upset by radiation. There are several units used to measure radiation effects on living matter. Until recently the most common have been the 'radiation absorbed dose', or 'rad', and the 'roentgen equivalent man', or 'rem'; the new international standard units are the 'gray' (Gy), equal to 100 rads, and the 'sievert' (Sv), equal to 100 rem. The rem and sievert allow for the greater severity of alpha or neutron damage for equivalent energy delivery. For beta and gamma radiation one gray is about the same as one sievert; for neutrons and alpha particles one gray may be up to 20 sieverts, depending on the energy of the particles.

The question of the biological effects of radiation is surrounded by controversy. But it is known that a dose of perhaps 400 rem of radiation over the whole body will kill half the adult human beings exposed to it; and very much smaller doses will produce cell damage that may lead to leukaemia and other kinds of cancer. Furthermore, radiation damage to the complex molecules in the reproductive cells which contain the hereditary information may produce mutant offspring. Even a single gamma ray can disrupt a gene; it may produce unforeseeable effects if this particular gene should be in a reproductive cell which subsequently helps to form a child.

A more detailed discussion of radiation biology is given in Appendix B. Suffice it to say here that the danger of radiation to living matter seems to increase in direct proportion to the amount of radiation exposure, beginning from the very lowest doses. There does not appear to be a threshold dose – that is, one below which damage does not occur. We are already subjected to continual radiation from the natural radioactive substances in our surroundings, and from cosmic rays. Any human activity which tends to add further sources of radiation to our surroundings must be potentially harmful. Just how harmful – and in return for what benefits – is still under debate; this book is intended to make one aspect of the debate more intelligible, whatever your criteria.

The Chain Reaction

In a lump of uranium there are always a few stray neutrons, produced either by spontaneous fission or by cosmic rays. Suppose that one of these stray neutrons induces a nucleus of uranium-235 to undergo fission. As well as the two fission products, there shoot outward perhaps two or three high-energy neutrons. (The chances are better than 99 to 1 that these neutrons will emerge virtually at the instant of fission: 'prompt' neutrons. But there is a slight chance that a neutron will not emerge until some seconds later: a 'delayed' neutron. As we shall see, delayed neutrons are of considerable importance.) There are three possibilities open to the high-energy neutrons from fission. A neutron may reach the surface of the material and escape. It may strike another nucleus and be absorbed without causing any immediate breakdown. Or – most importantly – it may strike another nucleus and, in turn, cause this nucleus to rupture. The chances of a neutron causing such an induced fission depend on the neutron's energy and on the nucleus it strikes. A fast neutron, fresh from an earlier fission, tends to go right through a nucleus so fast that nothing happens to the nucleus. Once in a while a fast neutron will rupture a nucleus; indeed only a fast neutron can rupture a nucleus of uranium-238. However, if a neutron ricochets among other nuclei, bouncing off each and giving up its energy bit by bit, it soon slows down until it is just jostling with the shared heat-energy of the rest of the material. It is then a 'thermal neutron'. A thermal neutron takes much longer to go through a nucleus, and is thus much more likely to rupture a nucleus of uranium-235 than is a fast neutron.

If, in a lump of uranium-235, one nucleus undergoes fission, the neutrons it releases may strike other nuclei, causing more fissions and releasing more neutrons. If there are enough uranium-235 nuclei sufficiently close together, the spreading disruption multiplies with astonishing rapidity: more and more neutrons, more and more ruptured nuclei, their fragments flying, more and more energy: a 'chain reaction'. If there is enough uranium-235, packed closely together for long enough, and if the chain reaction is out of control, the result is a nuclear explosion: an 'atomic bomb'. Slamming two suitable hemispheres of uranium-235 together at a very high velocity will indeed create a nuclear explosion; but there are other, much more efficient, techniques – and materials.

Needless to say, as soon as an appropriate arrangement of appropriate material became possible it was tried out – on 16 July 1945, at the top of a tall tower in the desert near Alamogordo, New Mexico: the world's first nuclear explosion, code-named Trinity. Within three weeks an atomic bomb made of uranium-235 devastated Hiroshima. But, seemingly, one 'doomsday weapon' was not enough, and uranium-235 was not the only nucleus that could be used. A neutron can penetrate a nucleus of uranium-238 without rupturing it. If this happens, the resulting neutron-heavy nucleus soon emits a beta particle, and then another, to become a nucleus of plutonium-239. Like uranium-235 – and only a few other isotopes, all at present very rare – plutonium-239 is 'fissile': that is, it can undergo a chain reaction of successive fissions, as the Trinity test demonstrated. On 9 August 1945 such a chain reaction obliterated Nagasaki.

The Nuclear Reactor

If chain reactions in uranium-235 and plutonium-239 could be used only in weapons, the situation would already be sufficiently complicated. But, more than two years before enough pure fissile material of either kind had been accumulated to make a weapon, it was found to be possible to control a chain reaction: to have it maintain itself without multiplying out of control. Indeed it was by this means that the plutonium was produced for the Alamogordo and Nagasaki bombs. The arrangement used to create and control a sustained nuclear chain reaction is called a 'nuclear reactor'.

The difference between an uncontrolled and a controlled chain reaction is profound. An arrangement of fissile nuclei which is to undergo an uncontrolled chain reaction – a nuclear explosion – must be sudden and final. An arrangement of fissile nuclei which is to sustain a continuing controlled chain reaction must be much more carefully organized. Curiously enough it takes many more nuclei – that is, much more material – to build a reactor than it takes to set off an explosion. This is of course partly because an explosion requires comparatively pure fissile material; in a reactor the fissile material is comparatively dilute, and there is accordingly much more material in all. But there must also be many more fissile nuclei themselves. The reason for this is the role played by the all-important neutrons.

If a chain reaction is to be self-sustaining, it must keep itself supplied with neutrons. Consider the following typical sequence. A neutron plunges into a nucleus of uranium-235. The nucleus ruptures; as well as two fission-product nuclei it also shoots out three neutrons. One of these three goes out through the surface of the lump of uranium and is lost. Another is absorbed by a nucleus of uranium-238, which begins its two-stage change into plutonium-239, but does not rupture. This leaves one neutron. If this third neutron now plunges into another uranium-235 nucleus and ruptures it, the process can continue; otherwise the chain reaction is snuffed out.

At any instant, inside the lump of uranium, there must be the right number of neutrons of the proper energy to propagate the chain. In effect, for a sustained chain reaction, each neutron which is lost by causing a fission must be replaced by exactly one neutron which does likewise. The system then has a 'reproduction factor' of 1. When this condition is achieved the system is said to be 'critical', and the situation is called 'criticality'. Despite a common misconception to the contrary, 'criticality' is not here used to imply 'danger'. You say that a nuclear system 'goes critical' just as you say that a car engine 'starts'. If on average each neutron lost when it causes a fission is replaced by more than one which also causes fission, the reaction 'runs away'; the reproduction factor is greater than 1, and the system is 'divergent'. If on average each neutron so lost is replaced by fewer than one which causes fission, the reaction will stop; the reproduction factor is less than 1. This is why a piece of uranium below a certain minimum size cannot under normal conditions sustain a chain reaction: there is too much surface through which neutrons can leak out.

Moderators

The basic requirements for a continuing controlled chain reaction are therefore, first, a collection of fissile nuclei appropriately distributed in space; and, second, a self-replenishing supply of neutrons of just sufficient numbers and energy to keep the chain reaction going. In natural uranium, only 0·7 per cent of the nuclei are fissile uranium-235. These fissile nuclei, only seven out of every thousand, are not sufficiently close together to keep up a chain reaction; too many neutrons are absorbed by the heavier uranium-238 nuclei, without causing fission. To improve the prospects for a sustained chain reaction it is necessary either to increase the proportion of uranium-235 relative to uranium-238; or to slow down the neutrons to thermal energies, at which they are much more readily absorbed by uranium-235; or to do both.

As we shall see (pp. 78–82), increasing the proportion of fissile uranium-235 – so called 'enrichment' of the uranium – is a complex and expensive process. But even a small increase, say from 0·7 per cent to 2 or 3 per cent, makes a marked difference, provided that the neutrons from fission are slowed down. This can be done by means of a material with light nuclei – a 'moderator'. A fast neutron striking a light nucleus in the moderator gives up a fraction of its energy, and after a few such collisions has slowed to thermal energy. The best moderators are the lightest nuclei: those of hydrogen. Ordinary water, containing two hydrogen atoms per molecule, is a satisfactory moderator. But ordinary hydrogen nuclei absorb neutrons. Better still is a rarer form of hydrogen nucleus: the proton-plus-neutron form called 'heavy hydrogen' or 'deuterium'. If two atoms of heavy hydrogen combine with an atom of oxygen, the result is a molecule of 'heavy water' or deuterium oxide (sometimes written D_2O), which is much the best moderator of fast neutrons.

One other substance is widely used as a moderator: carbon, in the form of graphite. A carbon nucleus – six protons and six neutrons – is much more massive than either form of hydrogen nucleus, and is therefore not such a good moderator. But graphite is less expensive than heavy water; furthermore it is a solid, which can be structurally useful in a reactor.

Reactor Design and Operation

To set up a nuclear reactor you proceed as follows. You take a good many pieces of material containing uranium-235 – usually uranium metal or oxide, natural or enriched: the 'fuel'. (You can also use plutonium-239, although this – as we shall see – involves some difficulties.) For a large reactor you need many tonnes of fuel, much more than enough to achieve criticality. One obvious reason for the extra fuel is to enable you to operate the reactor for some time before replacing the fuel. Other reasons will become clear in a moment.

You seal the pieces of fuel into casings called 'cladding', to support the fuel and to confine the fission products that will be produced. You position the assemblies of sealed fuel, called 'fuel elements', supporting them as necessary; remember that they may be very heavy indeed. You intersperse the fuel elements with moderator, to slow down the neutrons, and with neutron absorber, to enable you to control the chain reaction. You also include measuring instruments to tell you what is going on inside the reactor. You need to know, in particular, the temperature and the concentration of neutrons at various places inside the reactor.

You are now ready to start up your reactor. Before start-up, with all the absorbers in the interior of the reactor soaking up neutrons, the neutron density is so low it is difficult to measure, unless you intentionally include a separate source of neutrons as a sort of primer. A common form of absorber is a rod thrust through the interior of the assembly: a 'control rod'. Such a rod incorporates a material like boron, which absorbs neutrons like a sponge. The rod may be made, for instance, of boron steel. So long as enough control rods are in place no chain reaction is possible. To start up your reactor – to make a chain reaction possible – you begin withdrawing control rods.

The region of the reactor in which the reaction takes place is called the core. You withdraw control rods very slowly from the core, usually in short steps, in suitable symmetry to maintain a more or less uniform build-up of neutron density inside the reactor. In due course your reactor 'goes critical': a self-sustaining chain reaction is established, in which each neutron lost by causing a fission is replaced by exactly one neutron (either prompt or delayed) which does likewise. If the chain reaction could be sustained by prompt neutrons alone it would be 'prompt critical', and difficult to control. The dependence of the chain reaction on delayed neutrons allows you to adjust the reaction-rate gradually instead of abruptly.

Taking absorber out of a stable chain reaction is called 'adding reactivity'; the neutron density increases, and the rate of the chain reaction increases. But the build-up is gradual, because some of the neutrons do not emerge immediately after fission. The smaller the added reactivity, the longer is the time taken for the neutron density to increase by a given proportion. This time is called the 'reactor period', and is a very important measure of how well the reactor can be controlled. When a reactor has a short period it is liable to be skittish.

Of course inserting absorber – 'adding negative reactivity' – produces a reverse effect. When the desired rate is established you reposition the absorbers to stabilize the reaction at that rate.

To economize on neutrons you can surround the reactor core with a reflector to bounce errant neutrons back into the reaction region. The best reflecting materials are the moderator materials; in effect you can extend the volume of moderator beyond the region of fuel elements. Since the presence or absence of reflector affects the neutron density in the core, you can add reactivity by adding reflector, or vice versa; some reactor designs utilize this effect for control purposes.

Before pulling the control rods far enough out to let your reactor go critical you must take precautions against the radiation pouring out from the core. Neither alpha nor beta particles will get beyond the fuel cladding (unless it leaks); but gamma rays and neutrons can travel through metres of concrete and still be dangerous to living matter. Therefore, you surround your reactor with enough concrete or other protective 'shielding' to cut down the radiation outside to as low a level as you think advisable.

Xenon Poisoning

Normal start-up and shutdown of a reactor are both lengthy processes, and may take many hours. If it is necessary to stop the chain reaction quickly, for instance in the event of a malfunction, the emergency shutdown is called a 'scram'. If an operating reactor is left to itself its reaction-rate will gradually dwindle, not necessarily because fissile nuclei are being used up – in some reactors the number of fissile nuclei may even be increasing – but also because of the build-up of fission products which absorb neutrons.

The most voracious of all is xenon-135. The consequent phenomenon, called xenon poisoning, is an intriguing demonstration of the slightly surrealistic circumstances in which reactors operate.

When you start up your reactor for the first time, the fuel contains no xenon-135. For several hours after start-up, fission processes generate tellurium-135 and iodine-135, which in turn generate xenon-135, which starts gobbling neutrons. Each xenon-135 nucleus which succeeds in capturing a neutron is thereby changed into xenon-136 – much less voracious. The xenon-135 nuclei which fail to capture neutrons nonetheless undergo beta decay into caesium-135 – also much less voracious. Accordingly, after matters have had a chance to settle down, as much xenon-135 is being lost as is being generated. There is a certain average concentration of xenon-135 in the reactor core, which remains the same as long as the chain reaction proceeds at the same rate. You budget for so many neutrons lost to xenon-135, and operate accordingly. But when you change the reaction-rate you upset the balance, and the consequences may be embarrassing.

Iodine-135 turns into xenon-135 with a half-life of 6·7 hours. Xenon-135 turns into caesium-135 with a half-life of 9·2 hours – that is, slightly more slowly. Suppose you shut down your reactor. The neutron flux falls to near zero; xenon-135 stops capturing neutrons. From the moment of shutdown more xenon-135 is being generated than is being lost: while your reactor is shut down, the amount of neutron absorber in its core is steadily, surreptitiously increasing. If, several hours later, you try to start up your reactor again, you may, even with the control rods completely out, be unable to add enough reactivity to reach criticality. To be always able to start up your reactor at any time after shutdown, you will find it necessary to include more fuel, or otherwise to arrange for an excess of available reactivity over and above what normal operation needs. Apart from the obvious cost of the extra fuel, this means that even in normal operation you must leave some control rods partly inserted. It is not easy to do so without distorting the uniform neutron density in the core, and producing a less than ideal pattern of chain reaction. Reactor designers have to decide what sort of compromises they can best achieve, to satisfy the conflicting requirements made necessary by phenomena like xenon poisoning.

Refuelling

While you operate your reactor, changes take place in the fuel. The number of uranium-235 nuclei dwindles gradually as they undergo fission. Some of the uranium-238 nuclei capture neutrons and change to plutonium-239. Some of these plutonium-239 nuclei undergo fission. Others capture additional neutrons and become plutonium-240, plutonium-241 and other isotopes of elements heavier than uranium

– 'transuranic actinides'. Fission products are formed; most fission products are radioactive, and undergo radioactive changes into more stable nuclei, some very rapidly, others very slowly. Fission products also capture neutrons. The composition of the reactor fuel grows increasingly complex as the chain reaction proceeds; it becomes more and more difficult to keep track of all the competing processes taking place. Some of the fission products are gaseous, like krypton and xenon; these gaseous fission products build up inside the fuel, exerting pressure and trying to leak out. The intense neutron flux plays havoc with the crystal structure of the fuel, the cladding, and possibly the moderator, knocking the nuclei out of place and setting up stresses and strains in the material. Soon or later, it is necessary to take out used fuel and replace it.

There is an assortment of different procedures for 'refuelling' or 'recharging' a reactor. Some designs can be refuelled while the reactor is in operation, replacing one or more fuel elements at a time: 'on-load refuelling'. Other designs are shut down for refuelling, and perhaps one-third of the core is replaced at one time: 'off-load refuelling'. All refuelling procedures must be carried out with extreme care because of the intense radioactivity of the fission products in the reactor core and in the used, 'spent' or 'irradiated' fuel.

Power from a Reactor

If you set up a reactor on a sufficiently large scale, and let the chain reaction run fast enough, the energy released by the rupturing uranium-235 (and plutonium-239) nuclei makes the whole assembly hot – potentially very hot indeed. Complete fission of all the nuclei in a kilogram of uranium-235 would release a total energy of about one million kilowatt-days – that is, as much heat as would be given off by one million one-bar electric fires operating for one 24-hour day. That is a lot of heat. Accordingly, the fuel in a reactor must be arranged so that the heat is given off gradually enough to keep temperatures manageable. The amount of heat given off per unit volume in a reactor core is called the 'power density'. It may be anything up to several hundred kilowatts of heat per litre; if such an outpouring of energy is not to melt – and indeed boil – the whole aggregation of material, it must be efficiently removed.

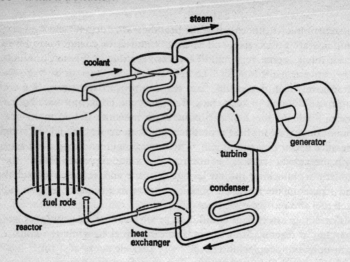

Figure 1   Nuclear power station

You remove the heat from the reactor by pumping a heat-absorbing fluid through the core, past the hot fuel elements. The fluid can be a gas, such as air, carbon dioxide or helium; or a liquid, such as water or molten metal. The choice of cooling fluid – 'coolant' – depends on how fast heat must be removed; on how expensive the fluid is; on how easy it is to pump; and so on. The cooling system can be open-ended, passing ordinary air or water directly through the core and back to the atmosphere or river; such an arrangement has the virtue of simplicity, but may also have serious drawbacks, especially if fuel cladding leaks. Alternatively the cooling system can be one or more closed circuits, in which the same coolant passes through the core again and again, carrying heat out of the core, discharging it outside the reactor, and then passing the rest of the way around the circuit and back through the core again. If the cooling system is made up of closed circuits, expensive or exotic coolants can be used, since they are confined in the system instead of being lost. A closed circuit can also be pressurized, which will in most instances dramatically improve efficiency of the coolant; a pressurized gas is denser and can carry more heat per unit volume.

The cooling system, of whatever design, removes the heat from the reactor core; what becomes of the heat thereafter depends on the reason for operating the reactor. The first large reactors were operated exclusively to generate neutrons and turn uranium-238 into plutonium-239 for nuclear weapons. The heat released in their cores was just a nuisance to be got rid of, into the near-by air or water. But with appropriate arrangements such heat, like the heat from burning coal or oil, can be used. In particular it can generate steam to run turbines or other electrical generators. Such an arrangement – a nuclear reactor providing heat to run an electrical generating plant – is called a nuclear power station, or (in the USA) a nuclear power plant.

In the following chapter we shall look much more closely at the structure and operation of the several main types of reactors. Each type releases energy by the fission of nuclei; but each uses a different arrangement of these nuclei, different designs of fuel, of moderator, of cooling system, of controls, and so on. The differences have many important implications, as we shall see.

2 · Reactor Types

With different fuels, moderators, control systems, cooling arrangements, spatial configurations and so on, possible designs of nuclear reactor number in the hundreds. Early reactor designers had a field day, letting their imaginations run riot; some of their suggestions made colleagues' hair stand on end. Others seemed more feasible: amenable to engineering, using manageable materials, controllable, safe, and – ultimately – even economic to build and operate.

As we shall see, the main lines of development of such commercial reactors sprang from the three partners in the Second World War 'atom-bomb' programme, the 'Manhattan Project'. The UK in due course developed gas-cooled, graphite-moderated reactors; the USA developed reactors cooled and moderated by ordinary 'light' water; and Canada developed reactors moderated by heavy water, variously cooled. Both the UK and the USA also began development of reactors using fast neutrons, with liquid metal coolant and no moderator. Before we describe these and other reactors in detail, it may be useful to identify some general aspects of reactor design.

To generate a given output of energy a reactor may have a very large volume of core, with a comparatively low heat output per unit volume or power density; alternatively it may have a much more compact core with a higher power density. Natural uranium reactor fuel has a low concentration of fissile nuclei; a reactor using such fuel must have a larger volume of core than one using enriched uranium – or plutonium – fuel. A large reactor costs more to build than a smaller one of the same output. On the other hand, natural uranium fuel is much cheaper than enriched uranium fuel. What you lose by building the larger, more expensive reactor you may subsequently save on fuel costs.

The energy output from a reactor can be measured directly as heat. If this heat is used in a 'power reactor' to generate electricity, only a fraction of the total heat energy ultimately reappears as electrical energy; the rest is discharged to the surroundings as low-temperature heat. In general, the higher the temperature the reactor can achieve, the larger the fraction of energy that can be converted to electricity. As a rule only some 25 to 32 per cent of the total heat output is converted to electricity in systems now operating. A system which converts 30 per cent of the heat to electricity is said to be 30 per cent efficient – mainly because the remaining 70 per cent of the heat is not used. (This is not to say that it cannot be used, merely that it is not.) Reactor energy outputs are accordingly described either as heat – for instance, 'megawatts thermal', MWt – or as electricity – for instance, 'megawatts electric', MWe. (A megawatt is one million watts.) A satisfactory rule of thumb is to assume that, for a given power reactor, the output in MWe is between one quarter and one third of the output in MWt. Unless use is made of the low-temperature heat, the fraction MWe/MWt is a measure of the system's efficiency.

If a reactor core operates at a higher temperature, it produces steam of higher quality, and generates electricity more efficiently. On the other hand, core materials which withstand these higher temperatures are likely to be more expensive. Similarly, reactor fuel which can be left in the core for a longer period at a higher temperature reduces the amount of fuel needed for refuelling; but such fuel also costs more. A reactor which can be refuelled 'on load' – without having to shut down – is less inconvenient for an electricity system; but such refuelling arrangements are in general more expensive to build than those for 'off-load' refuelling.

The cooling system of a reactor may operate at a pressure anywhere from atmospheric up to – at present – about 150 atmospheres. The higher the pressure, the heavier and stronger must be the pressure system. This has implications not only for costs but also for safety, since a rupture of the pressure system might have serious consequences, as we shall see. Some designs enclose the reactor core in a pressure vessel of heavy welded steel; other use pre-stressed concrete. Still others distribute the core materials in an array of much smaller pressure tubes.

Interruption of cooling may be easier to control in a reactor of low power density than in one of high power density, in which sharp temperature rises may occur with extreme rapidity. Flaws or malfunctions in the pressure system may be easier to overcome in a reactor with low coolant pressure than in one with high coolant pressure; a large welded steel pressure vessel of complex geometry seems inherently more vulnerable to major disruption than a pressure vessel of pre-stressed concrete, or a system made up of many smaller pressure tubes.

One large reactor may cost less than two small ones producing the same total output – but, as we shall discuss (pp. 181–2), not necessarily, if the large reactor must add many extra items of equipment for reasons of safety, stand-by and maintenance. All reactor designs have shared one characteristic: a rapid increase in the size of successive reactors of the same basic design. In reactors, perhaps more than in most other engineering technologies, a change of scale is frequently not merely quantitative but qualitative, introducing a whole new set of unknowns into the engineering. In the table on pages 72–3 we have listed typical design parameters of different reactor types. In the succeeding sections we shall describe them in more detail.

Experimental and Research Reactors

The first nuclear reactor was constructed amid tight wartime secrecy, in a disused squash court under Stagg Field football stadium at the University of Chicago. Construction of the reactor began in November 1942 and took less than a month. Bricks were machined out of graphite. In some of the bricks were embedded balls of uranium metal or compressed uranium oxide powder; uranium oxide had to be used because at that time only 5600 kg of pure uranium metal were available. The graphite bricks were laid layer after layer onto a growing pile of roughly spherical configuration inside a wooden supporting structure. At intervals inside the pile were neutron-absorbing cadmium strips to ensure that stray neutrons did not initiate a premature chain reaction. Instruments to measure neutron-density were also included, and checked regularly to see how the pile was progressing towards critical dimensions.

By the time the 57th layer of bricks had been added it was clear that only the inserted neutron absorbers were keeping the pile from criticality. By this time it was more than 6 metres high, with length and breadth to match, and contained about 36 tonnes of uranium and over 340 tonnes of graphite.

On 2 December 1942 the scientists and technicians gathered on the balcony of the squash court, watching the instrument readings, while Enrico Fermi called out instructions and a young physicist named George Weil slowly pulled out the final control rod. Shortly after 2.30 p.m. the instruments recorded a steadily rising increase in neutron density in the pile. The pile had 'gone critical': the first self-sustaining nuclear chain reaction was taking place.

The heat generated in the pile was initially kept down to about 0·5 watts. But on 12 December the reaction-rate was allowed to increase until the heat generated – the 'power level' – reached 200 watts. Further reactivity was available, but by this power level the radiation from the pile was potentially harmful to personnel. Accordingly, in the spring of 1943, Chicago Pile No. 1 – CP-1, as it came to be called – was unpiled. Shortly thereafter, rebuilt with added uranium and graphite inside adequate radiation shielding, at a site outside Chicago, and rechristened Chicago Pile No. 2, it could be operated at an average power of 2 kilowatts (2 kWt) and intermittently up to 100 kWt. For some years, until the name became totally inappropriate, any nuclear reactor was called an 'atomic pile', after the first reactor.

CP-1 was the first true nuclear-reactor. But even before it was built, thirty piles of less than the necessary size and shape were built and tested. Such assemblies, which cannot generate their own neutron supply without a supplementary source of neutrons, are called 'sub-critical assemblies'. Since the early 1940s countless hundreds of sub-critical assemblies have been built and dismantled in many countries; and many true reactors have been built for experimental or research purposes. The variety and range of designs of experimental and research reactors is extensive, depending on the purpose for which a particular reactor is constructed. Experimental and research reactors have a number of uses. To bombard a sample of material with neutrons, the sample can be inserted through a suitable channel into a reactor core. The intention may be simply to study the effect of neutron bombardment on the material – perhaps a material to be used in building a reactor. Or the intention may be to convert some of the sample's stable nuclei by absorption of neutrons into radioisotopes, for medical, industrial, agricultural or research purposes. Some reactors have a 'thermal column': a graphite panel through the reactor shielding, which allows a stream of thermal neutrons to emerge for research work outside the reactor. (Although thermal neutrons are often called 'slow neutrons', their speed is nonetheless about 2200 metres per second – considerably faster than a high-velocity bullet.)

Some research reactors are designed to further the study of 'reactor physics' itself: neutron densities, temperatures, the production of plutonium-239 from uranium-238, the build-up of fission products, the effect of these fission products on reactivity, the performance of new designs of fuel assemblies, the effects of 'unscheduled events' inside the reactor, and so on.

Obviously, research reactors are also important for the training of qualified scientists and technicians in the often extremely subtle and intricate – and potentially dangerous – characteristics of reactor design and operation. Many countries now boast major centres for reactor research and development. Nor are such reactors found only in heavily industrialized countries. One of the longest-serving research reactors in the world, in operation since 1959, is the 1-MW Trico reactor in Zaire.

A popular design of research reactor is the 'pool type': it has a core of highly enriched uranium at the bottom of a deep tank of water. The water acts as moderator, reflector, coolant and shielding. It also allows a direct view of the core while the reactor is critical; in no other reactor design is this possible. Because some of the radioactive emissions from the reactor travel faster than the speed of light in water, the water in a pool-type reactor glows with an eerie blue light called Cerenkov radiation.

The great majority of reactors, of whatever size or for whatever purpose, have been experimental, in that every new reactor modification and development has had to be designed and engineered on the basis of previous experience, which in this field is often inadequate or irrelevant or both. The US Atomic Energy Commission (see p. 109) went so far as to list all the reactors it licensed, for whatever purpose, as 'experimental' until 1971.

Plutonium Production Reactors

All uranium reactors produce plutonium, by neutron bombardment of uranium-238. The first large-scale reactors were built expressly for this purpose: to produce plutonium for nuclear weapons. A pilot model was built in 1943 at Oak Ridge, Tennessee. It could not be constructed on the simple building-block principle that sufficed for CP-1; it would have been not a little inconvenient to dismantle the entire reactor to recover the plutonium. Furthermore, the rate of transmutation of uranium into plutonium depends on the neutron density, which in turn depends on the rate of the chain reaction. If the reaction is fast enough to create plutonium at a useful rate the heat generated becomes a major problem. Complete fission of all the nuclei in one kilogram of uranium-235 releases about one million kilowatt-days of energy; each uranium-235 fission is likely at most to initiate one further fission with one neutron and create one uranium-239 (and hence plutonium-239) nucleus with another. That is, to create one kilogram of plutonium-239 requires the fission of about one kilogram of uranium-235 – and the dissipation of all that heat.

Accordingly, the Oak Ridge reactor was built in the form of a cube of graphite perforated from one side to the other with parallel horizontal channels. Into these channels were slid cylindrical slugs of natural uranium clad in aluminium. When a fuel slug had been sufficiently irradiated it was pushed through the reactor, falling out of the graphite core and into a tank of water, for subsequent processing (see pp. 89–90). The fuel slugs fitted loosely in the channels, leaving room for a flow of cooling air to remove the heat from the reaction (eventually 3·8 MWt).

Even while the Oak Ridge pilot model was still under construction, work began on the first full-scale reactor, which was built on the bank of the Columbia river near the town of Richland, in Washington state. Construction of the first full-scale reactor, as tall as a five-storey building, only took from June 1943 until September 1944. By early 1945 three full-scale reactors were in operation. The entire industrial installation, named the Hanford reservation, was in due course to occupy nearly 1600 square kilometres, and include nine production reactors, plus a vast array of ancillary plant. The Hanford production reactors were similar in design to the Oak Ridge reactor; but their heat output was so intense that cooling by gas – helium was the original choice – was found too difficult. Cooling was accomplished by pumping water from the Columbia river directly through a reactor core and back into the river.

After the end of the Second World War plutonium production reactors were built in the UK, France and the Soviet Union. The UK production reactors were built on the Cumbria coast, on the site of a disused ordnance factory which was renamed Windscale. Like the Hanford reactors those at Windscale used natural uranium clad in aluminium, lying in horizontal channels in a graphite core. The absence of a suitable water supply meant that the Windscale reactors were cooled with air, blown by powerful fans through the cooling channels in the graphite, and discharged through a stack 126 metres tall back to the atmosphere. This once-through air cooling was a far from desirable arrangement, whose drawback was subsequently demonstrated in the dramatic accident in 1957 which destroyed the Windscale No. 1 reactor (see pp. 123–6).

If the purpose of a reactor is to produce fissile plutonium-239, the rate of plutonium production can be 'optimized' by choice of core geometry, at the cost of other performance characteristics. The reactor fuel must be changed at relatively short intervals – months rather than years. By this time the fissile plutonium-239 in the fuel is playing a significant part in the chain reaction, undergoing fission and thus being removed as well as formed. Furthermore, some of the plutonium-239 absorbs one or more additional neutrons without undergoing fission, becoming plutonium-240, plutonium-241, and plutonium-242. Plutonium-240, which accumulates comparatively rapidly, is susceptible to spontaneous fission, but is unlikely to fission when struck by a neutron, and cannot therefore participate in a chain reaction. It is moreover virtually impossible to separate from plutonium-239 (but see pp. 208–9). Too high a fraction of the 240 isotope makes plutonium somewhat unpredictable as a weapons material: hence the need to remove irradiated fuel before too much plutonium-240 has been created. However, 'reprocessing' of fuel to extract plutonium (see pp. 86–9) is an expensive and complex operation. Refuelling more often than is strictly necessary to maintain a reactor's reactivity can only be justified within the remarkable elasticity of military budgeting.

Gas-cooled Power Reactors

Magnox Reactors

The first power reactors were of course, like the plutonium production reactors, military: power plants for submarines, and multi-purpose reactors producing both plutonium and electricity. (A 'nuclear submarine' is so called as much for its motive power as for its cargo.) The first 'power reactors' so identified were started up in the USA and the Soviet Union in 1954. The US reactor had an output of 2·40 MWe, and the Soviet APS-1 reactor at Obninsk, now usually declared to have been the world's first power reactor, an output of 5 MWe.

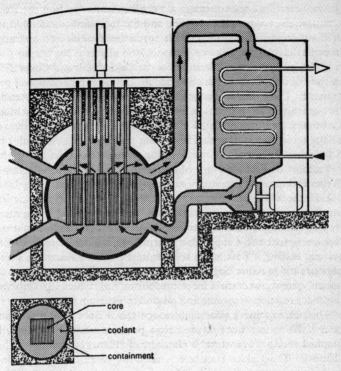

core
coolant
containment

Figure 2   Magnox reactor

However, for obvious reasons, the general public heard little about the first US and Soviet power reactors. By default, if not by common consent, the world's 'first nuclear power station' was Calder Hall in the UK, whose first reactor started up in 1956. Calder Hall's claim to precedence is entirely defensible, if only because the first Calder Hall reactor, like its three successors, was a full order of magnitude larger than the Obninsk reactor, with an output of 50 MWe. On 17 October 1956 Her Majesty Queen Elizabeth II switched power from Calder Hall into the UK's National Grid: in a blaze of international publicity the age of 'nuclear power' – that is, electricity, not military might – was born.

The four Calder Hall reactors, on a site adjoining Windscale, were 'power' reactors only secondarily. Despite the fanfare and the Royal premiere the Calder Hall reactors, and the four similar reactors built at Chapelcross across the Scottish border, were built and optimized in order to produce weapons-plutonium to augment the output from the Windscale reactors. Nonetheless the Calder Hall and Chapelcross nuclear stations became the cornerstone of the UK nuclear power programme. Their design characteristics were developed and extended through the first generation of UK commercial nuclear stations, eventually comprising a total of twenty-eight reactors, including one in Italy and one in Japan. The nuclear patriarch of this family, the first Calder Hall reactor, is still going strong more than twenty-five years after its first start-up. Many of the factors which affected its design and construction still preoccupy nuclear engineers.

Like the Windscale reactors this Calder Hall reactor uses natural uranium fuel and graphite moderator. But their spatial arrangement is very different, as are many other details. The fundamental difference is that the Calder Hall reactor has a closed-circuit cooling system, making it possible to recover heat from the reactor at a temperature and pressure high enough to be useful. The pressurized closed-circuit system also ensures more efficient cooling, which in turn allows the chain reaction to operate and produce plutonium faster.

The heart of the Calder Hall design (see Figure 2, and the table on pp. 72–3) is a huge welded steel pressure vessel, enclosing the graphite reactor core which is pierced from the top to bottom by fuel channels. The Calder Hall fuel is clad, not in aluminium, but in a special magnesium alloy called 'Magnox', which is much less inclined to absorb neutrons, and is stronger and less susceptible to corrosion in the high temperature and neutron flux inside the reactor core. The entire family of reactors using such fuel has always been referred to as Magnox reactors.

The core contains an array of instruments which transmit readings of temperatures, neutron densities and other relevant data to the control room. Each sector of the core also has channels for several types of control rods which enter the reactor from above, held out on electromagnetic grapples, so that any reactor fault will shut off the magnets and let the rods fall into the core to halt the fission reaction.

The pressure vessel, its contents and its attachments expand and contract with temperature changes. The combination of the resulting thermal stresses, the gravitational stresses set up by the weight of the components, the vibration of moving parts and fast-flowing coolant, and the somewhat unpredictable effects of prolonged intense neutron irradiation presented the Calder Hall designers with a challenge whose equivalent still faces every nuclear engineer. The steel pressure vessel is itself enclosed inside a biological shield of concrete more than two metres thick. Assorted pipes and services pass through the biological shield; but, since the penetrating gamma rays and neutrons travel in straight lines, an appropriate arrangement of zigzags cuts off all outcoming radiation. The total weight of the reactor and its ancillary structures is considerable – some 22 000 tonnes – and the site requirements are stringent; any subsidence might crack the concrete, reducing the effectiveness of the shielding.

The hot coolant gas passes out of the reactor building through four cooling ducts into four towering 'heat exchangers' – a fancy word for boilers. Inside each is a labyrinth of tubing containing water; the hot carbon dioxide passes round the tubing, giving up its heat to the water which turns to steam and is used to drive turbogenerators. When the gas has given up its useful heat it emerges from the lower end of the heat exchanger, and passes into a gas circulator. This blows it back into the bottom of the reactor pressure vessel and up again through the fuel channels. Since the four loops of the cooling circuit are pressurized, special provisions have to be made for changing fuel elements and for other maintenance inside the reactor core. Access to the channels for refuelling and servicing is from above, through holes in the horizontal roof of concrete shielding which is called the 'pile cap'. On the pile cap, the working area above the reactor, are mobile 'charging' or 'refuelling' machines, massive and complex assemblies.

The Calder Hall Magnox design – primarily for plutonium production – is shut down and depressurized for refuelling. In the later Magnox designs for commercial nuclear stations it is not necessary to interrupt the operation of the reactor for refuelling; it can be carried out continually, a few channels per week, while the reactor is supplying power, 'on load'.

To change the fuel in the reactor, the 'discharge machine' is positioned over an access port, clamped onto the surface of the pile cap and pressurized. The shielding plug is removed, grapples extended down through a standpipe into the core, and the irradiated fuel elements lifted out of a channel and stored inside the thick walls of the discharge machine – all by remote control, because of the radiation hazards. The shielding plug is replaced, the discharge machine depressurized and moved, and the charge machine, loaded with fresh fuel, moved into position. The whole cycle is repeated, again by remote control, to lower new elements into place: clamping, pressurization, unplugging, replugging, depressurizing and unclamping.

In either case, the irradiated fuel, intensely radioactive with fission products, is moved, inside the discharge machine, to be dropped into a 'cooling pond': a deep tank of water, which serves to shield and cool the fuel while the more short-lived fission products within it decay to a less dangerous level of activity. After a suitable interval – usually 150 days – the irradiated spent fuel is transported to Windscale for 'reprocessing' (see p. 86ff.).

In all, eight Magnox stations, each with two identical reactors, were built for the Central Electricity Generating Board (CEGB), and one for the South of Scotland Electricity Board. Design details varied considerably from station to station, although all incorporated on-load refuelling by means of a single charge–discharge machine. The Berkeley station's reactors use cylindrical steel pressure vessels, whereas the Bradwell station's reactors, built at the same time, use spherical ones. The Hunterston station is refuelled not from above but from below, where the temperature is lower. The different stations have different arrangements of heat exchangers and generating sets, different reactor buildings and so on. Instead of a water-filled cooling pond for its spent fuel the Wylfa station has three gas-filled storage magazines (see p. 89). Perhaps the most important variation in the Magnox stations is the power rating, which was increased progressively. To accommodate this increase in power and size, the last two CEGB stations introduced a major modification in design. Welding of a steel pressure vessel of more than a certain size to the stringently high standards necessary for a reactor becomes prohibitively difficult. Accordingly, the CEGB's Oldbury station embodied an entirely new approach. The pressure vessel was fabricated not from welded steel but from pre-stressed concrete, a much more manageable material for large and complex structures. In the Oldbury design not only the reactor core but also the heat exchangers and gas circulators are enclosed within the concrete pressure vessel. The pre-stressed concrete serves both as pressure vessel and as biological shield; the gas ducts are completely eliminated, removing one of the major escape routes for radioactivity in the event of an accident. The pre-stressed concrete design made possible a more than twofold increase in reactor size for the final CEGB Magnox station, at Wylfa in Wales.

The power density of the Magnox stations, which averages about 0·9 kilowatts per litre, is, by nuclear standards, low. Because of its low density and consequent low thermal capacity, gas is a less efficient coolant than liquid; accordingly, the rate of heat generation in a gas-cooled core must be kept low. (This in turn imposes an overall limit on maximum feasible heat output since high output entails a very large volume of core, with accompanying engineering complications.) Another characteristic of interest is the 'specific power': power generated per unit mass of fuel. The specific power of Calder Hall fuel is about 2·40 kilowatts per kilogram of uranium; that of Wylfa fuel is about 3·16 kilowatts per kilogram of uranium. Specific power is also sometimes called 'fuel rating'. The 'burn-up' of fuel is the cumulative heat output per unit mass; it is commonly measured in megawatt-days per tonne of uranium. The burn-up is of course a measure of how many fissions have occurred inside a given amount of fuel.

One of the main objectives of fuel designers is to achieve higher burn-up – that is, to be able to leave fuel in a reactor longer, before it becomes too distorted and too burdened with fission products to function properly. The limitations on burn-up of Magnox fuel are numerous. Natural uranium metal has a complicated crystal structure, and undergoes a variety of unwelcome changes at high temperatures and intense neutron fluxes. A burn-up of about 5000 megawatt-days per tonne of uranium is about the best that can be comfortably attained by Magnox fuel. This limitation was one of several factors which eventually terminated the Magnox programme and provoked a search for another approach.

The only other major nuclear power programme to opt for gas-cooled reactors was the French. The small French power reactors at Marcoule and Avoine started up in 1958. The second unit at Avoine – Chinon-2, a 200-MWe reactor – went critical in 1964, and France thereafter built, in all, seven gas-cooled power reactors with graphite moderator and a 70-MWe gas-cooled reactor with heavy water moderator. But French interest then swung abruptly away from gas-cooled to light water designs, in partnership with American reactor-builders (see p. 140).

Advanced Gas-cooled Reactors (AGRs)

Even while the first Magnox stations were barely under construction, work commenced on a second-generation design of gas-cooled power reactor: the advanced gas-cooled reactor (AGR). The aim was to achieve higher gas temperatures to improve the efficiency of electricity generation; higher fuel ratings to make the reactor more compact; and higher burn-up to reduce the frequency of refuelling. The temperatures achievable with Magnox fuel are limited by the characteristics of Magnox alloy and of uranium metal. Uranium metal undergoes a crystalline change of phase at $665°$ C, accompanied by marked expansion; its behaviour even below this temperature is complex, since it expands at different rates in different directions with increasing temperature. The melting point of Magnox is about $645°C$; as well as melting at this temperature Magnox may also catch fire.

Accordingly, higher-temperature fuel must use some other form of uranium. The form most commonly chosen is uranium dioxide, UO_2, often just called uranium oxide. Whereas uranium metal melts at $1130°C$, uranium oxide melts only at $2800°C$. However, uranium oxide has a low thermal conductivity, much lower than that of uranium metal. When uranium metal is undergoing a fission reaction, its high thermal conductivity means that the temperature is more or less uniform through the whole thickness of a fuel rod, even if this is several centimetres. The same is not true of uranium oxide. If solid uranium oxide is undergoing fission, the heat generated in the interior does not readily make its way to the surface; the interior is much hotter than the surface. Uranium oxide fuel elements must have a smaller diameter than metallic uranium elements, even though the melting point of uranium dioxide is so much higher.

The basic building block of uranium oxide fuel is usually a pellet made by compressing, baking or otherwise persuading uranium oxide powder to assume the form of a small hard cylinder, about the size of a liquorice allsort. A column of such pellets, anything up to several metres in length depending on the fuel design, is stacked inside a thin-walled metal tube, to make a 'fuel pin'. The tube must be of a material which can withstand high temperature. Some oxide fuels use an alloy of zirconium called – unsurprisingly – zircaloy, which has advantages but is expensive; the usual alternative is stainless steel, as is the case with fuel for the advanced gas-cooled reactor. Stainless steel involves a further problem; it is strong and well-behaved structurally, but it has an unhealthy appetite for neutrons. Accordingly, the percentage of uranium-235 in the uranium oxide must be increased above its natural level: that is, the uranium oxide must be enriched (see pp. 78–82). In AGR fuel the uranium is usually enriched to about 2 per cent.

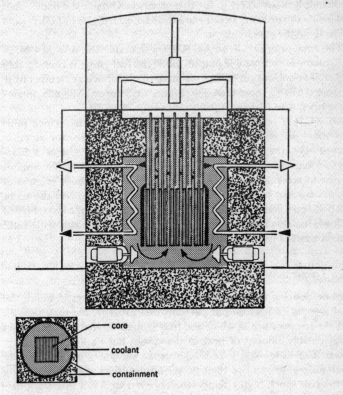

core

coolant

containment

Figure 3   Advanced gas-cooled reactor (AGR)

The first AGR incorporating this type of fuel was a small 28-MWe prototype built at Windscale, which started up in 1962. It was followed in due course, and after many misadventures, by a programme of five full-scale twin-reactor stations, and then by two more twin-reactor stations (see pp. 165–6).

The basis of the overall AGR design is the pre-stressed concrete pressure vessel first developed for the last two Magnox stations (see Figure 3, and the table on pp. 72–3). Like a Magnox reactor, an AGR has a core of machined graphite, under a dome like a huge steel bell-jar, with a large number of openings at the top, through which pass the standpipes for access to the fuel channels. Outside the dome – but still inside the pressure vessel – are the heat exchangers or boilers, and below them the gas circulators.

The amount of fuel in an AGR is considerably less than in a Magnox reactor of comparable output, while the fuel rating is considerably higher. The coolant gas emerges from the fuel channels at a temperature of around 650°C, more than 300° higher than normal Magnox operating temperature.

An AGR is refuelled by a single refuelling machine, which pulls an entire fuel string of eight elements out of the reactor at once. Accordingly, the refuelling machine is itself the height of a four-storey building, and the reactor building must be built like an aircraft hangar to accommodate it. A single machine serves both reactors at a given station, moving between them, the fuel store and the spent fuel pond on a gantry or rails. The AGRs were designed to be refuelled on load; but this has been one of many technical problems which still remain short of fully satisfactory resolution.

High Temperature Gas-cooled Reactors (HTGRs)

Anyone desiring a heat source will have two objectives in mind: the total amount of heat output per unit time (that is, the total power), and the temperature at which the heat is made available. There is an unimaginable amount of heat in the ocean; but its low temperature makes it of little overt use. While reactor designers were scaling up reactor sizes, to increase their power output, they were also pressing on towards much higher temperatures. Even an AGR operating flat-out is only a so-so source of heat, as far as temperature is concerned. It can be used to raise passable steam to run a turbogenerator and produce electricity, but only at a moderate efficiency. More elegant industrial applications are ruled out by the low temperature of the heat.

The limitation on the temperature at which heat is generated is nothing – or almost nothing – to do with the chain reaction system itself. Under the right circumstances a chain reaction can run at tem-

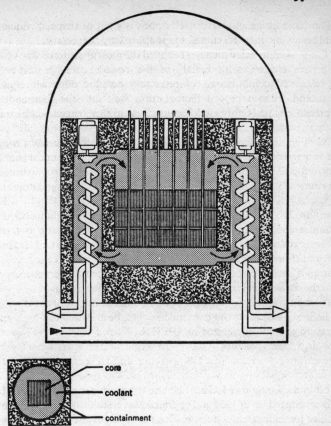

Figure 4   High-temperature gas-cooled reactor (HTGR)

peratures anywhere up to those in the heart of a nuclear explosion – millions of times higher than those in fossil-fuel boilers. However, long before such temperatures are reached it becomes peculiarly difficult to keep the whole assemblage in any semblance of order. We have already noted the awkwardness of uranium metal and Magnox cladding at temperatures over 600°C; other reactor materials present similar problems, albeit at variously higher temperatures. Clearly if really high temperatures are to be permissible, without having the entire reactor core bulge and warp itself hopelessly out of shape, or undergo unpleasant chemical reactions, a new approach is necessary.

The new approach which has received the most attention is one which dispenses entirely with metals in the reactor core, in favour of sophisticated combinations of refractory ceramic materials able to withstand without protest temperatures well into the thousands of degrees Celsius. Highly enriched uranium in tiny particles is blended intimately with the ceramic. Some designs proposed include also particles of another element called thorium. Thorium has nuclear properties rather like those of uranium-238. Natural thorium is almost entirely thorium-232. A nucleus of thorium-232 can absorb a neutron to become thorium-233, which then emits two beta particles to become uranium-233, which is fissile. The process is directly analogous to that by which uranium-238 is transformed into plutonium-239. Uranium-233, like uranium-235, undergoes fission when struck by a slow neutron and in turn produces more neutrons, to sustain a chain reaction. Uranium-238 and thorium-232, although not fissile materials, are called 'fertile' materials, because they can be transmuted by neutron bombardment into the fissile materials plutonium-239 and uranium-233.

All the various approaches to such a design have involved the choice of helium as coolant; the concept has thus been called the high temperature gas-cooled reactor or HTGR. Work began in 1957 on the first two such reactors. Under the aegis of the forerunner of the Organization for Economic Co-operation and Development (OECD), an international project was established at Winfrith, Dorset, England, to build the Dragon HTGR. The Dragon reactor, generating 20 MW of heat, started up in 1964 and operated successfully as an experimental facility for more than a decade. But the UK government, which was paying the majority of the programme's costs, decided that the HTGR design had no future in the UK. The Dragon breathed its last in 1976, slain by having its funds cut off. In the USA the General Atomic Company built Peach Bottom 1, the world's first HTGR power station, near Philadelphia. It started up in 1965. In the Federal Republic of Germany another small HTGR, the AVR, started up near Jülich in 1966. A larger prototype station, Fort St Vrain, Colorado, went critical in 1974. However, Fort St Vrain suffered seemingly endless teething troubles; Peach Bottom 1 was shut down permanently in 1974; and a radically different design of HTGR, the Thorium High Temperature Reactor at Schmehausen in the Federal

Republic, was at the end of 1982 likely to be abandoned for lack of further funds. The HTGR, once promoted as the most versatile and safest of all reactor designs, appears to have fallen at least temporarily by the nuclear wayside. Its possible advantages – increased efficiency of electricity generation, and even the production of nuclear heat for industrial processes like steam-making – now seem to be only of academic interest for the foreseeable future.

Light Water Reactors

Pressurized Water Reactors (PWRs)

Like the first British power reactors, which were built to produce weapons-plutonium, the first US power reactors also began under military auspices, albeit specifically as power plants. The US Navy realized after the Second World War that a submarine powered by nuclear fuel would not need to resurface to replenish oxygen supply, since the 'burning' of nuclear fuel – unlike that of oil – does not require oxygen. Spurred by this idea, and constrained by the space limitations in a submarine, US designers developed a reactor using a core of relatively high power density, with fuel elements immersed in a tank of ordinary water – called 'light water' to distinguish it from heavy water – under sufficient pressure to keep it from boiling. The 'first power reactor ever built', according to its builders, went critical on 30 March 1953 in a land-based mock-up of a submarine hull at the National Reactor Testing Station in Idaho. The following year saw the launching of the *USS Nautilus*, the first nuclear-powered submarine, powered by a pressurized water reactor system. In 1957 the submarine reactor came ashore, at the Shippingpoint power station near Philadelphia, the first nuclear power station in the USA. In subsequent years the pressurized water reactor, or PWR, has become the world's most popular.

The basic structure of a PWR (see Figure 5, and the table on pp. 72–3) is a large pressure vessel of welded steel with a lid held onto the upper end by a ring of heavy bolts. The pressure vessel contains the reactor core, and other so-called 'reactor internals' like control rods; the remaining volume is completely occupied by ordinary 'light' water under a pressure of about 150 atmospheres. The core is made up of fuel elements, each a faggot of 4m-long fuel pins. A PWR fuel pin is a tube of zircaloy, about 1 centimetre in diameter, filled with stubby cylindrical pellets of uranium dioxide. So far as neutrons are concerned the zircaloy cladding is comparatively well-behaved – much more so than stainless steel – albeit more expensive. But the water in which the whole concatenation is immersed is – as noted in

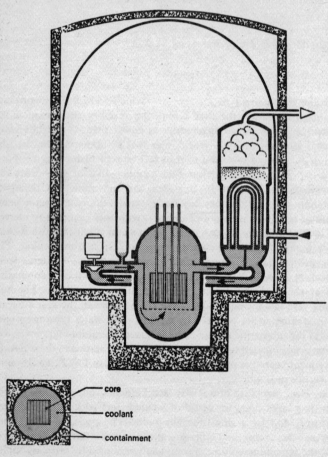

Figure 5   Pressurized water reactor (PWR)

Chapter 2 (p. 33) – an enthusiastic gobbler of neutrons, and to offset its distracting influence the uranium in PWR fuel pellets is enriched to upwards of 3 per cent uranium-235. The water inside the pressure vessel serves simultaneously as moderator, reflector and coolant. At the top of the core it leaves through heavy pipes welded to the pressure vessel. PWRs can have two or more 'loops' of cooling circuit. In each loop, the pipe through which the water enters the pressure vessel is called the 'cold leg', and that through which it leaves is called the 'hot leg'.

The hot leg of a PWR cooling loop carries the hot coolant water into a steam generator or boiler. The hot high-pressure water from the reactor passes through thousands of tubes immersed in more water, under considerably lower pressure. Although the pressurized water inside the tubes cannot boil, the lower-pressure water outside them does. The resulting steam is processed and piped to a turbogenerator set. The primary coolant water returns through the cold leg to the reactor vessel, encouraged by a primary coolant pump. One coolant loop also includes a 'pressurizer', in which an appropriate quantity of the coolant water is evaporated or condensed, to maintain coolant pressure and to compensate for the effects of thermal expansion and contraction as plant output varies. The pressurizer can also help to offset unintended increases in system pressure resulting from malfunctions. The electric immersion heaters in a pressurizer can generate 2000 kW – a bit overwhelming for a household hot-water system.

Easily the most controversial feature of the PWR are the emergency core-cooling systems, provided to prevent overheating of the reactor core in the event of an accident. However, rather than describing them here, it will be more appropriate to defer their description to Chapter 6 (pp. 141–3); there can be few technologies which have been subjected to such exhaustive – and inconclusive – scrutiny.

PWR control and instrumentation systems vary widely in design. But control rod assemblies are commonly suspended above the core, inside the pressure vessel lid, with drive mechanisms functioning through the lid from above. A PWR is refuelled off load – that is, with the reactor shut down. The reactor is allowed to cool. Then a pool-shaped chamber above the reactor – the 'reactor well' – is flooded with water, to provide shielding and cooling; the lid is unbolted and moved to one side, exposing the interior of the reactor. Since the whole procedure is time-consuming, a substantial proportion of the fuel charge is changed at each refuelling – typically about one third of the core. PWR designers usually provide for one refuelling operation annually.

Needless to say a PWR is, like any power reactor, enclosed in heavy shielding. The reactor vessel itself is surrounded by two or more metres of concrete, extending upwards to form the side walls of the reactor well. The concrete also encloses the entire primary circuit – steam generators, primary pumps, pressurizer and piping – because the primary coolant is commonly slightly radioactive (see pp. 90–92). The reactor building itself is usually designed to serve as a secondary containment.

Some PWRs deliver close to 4000 MW of heat at a power density over 100 KW per litre. But the low coolant temperature attainable using water under manageable pressure – some 150 atmospheres, as noted – makes the PWR a comparatively inefficient source of heat for electricity generation. Nonetheless PWRs greatly outnumber all other types of power reactor.

Boiling Water Reactors (BWRs)

US interest in water cooling of reactors stemmed from the Hanford reactors and was furthered by the submarine PWRs. It was known that water allowed to boil is more effective in removing heat, but boiling was thought likely to trigger instabilities in a reactor core. The water in such a core serves also as moderator; if a steam bubble forms, the local effect on reactivity is swift and its consequences difficult to predict. But experiments in the mid 1950s demonstrated that water could indeed be allowed to boil in a reactor core. Accordingly, a new design of reactor was developed, which is by far the simplest in concept of all power reactors: the boiling water reactor, or BWR (see Figure 6, and the table on pp. 72–3).

BWRs and PWRs are often mentioned in the same breath, as 'light water reactors' or LWRs. In a BWR the water serves as moderator, reflector and coolant – and in addition, when boiled, produces steam which is ducted directly to drive a turbogenerator. Once through the turbines, the coolant water is condensed and pumped again into the 'boiler' – that is, the reactor pressure vessel.

The pressure which the vessel must contain need not be much more than the pressure of the steam being produced – usually less than half

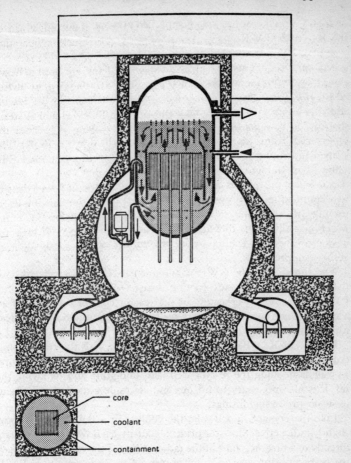

core

coolant

containment

Figure 6 Boiling water reactor (BWR)

that in a PWR. Accordingly, the pressure vessel need not be so thick. A BWR pressure vessel also includes the whole steam collection and processing array, above the core. The control rods therefore enter a BWR core from below. The cooling circuits of a BWR bear little resemblance to those of a PWR. In a BWR water boils inside the fuel assemblies, and there are no external steam generators. The consequent

saving in capital cost has long been billed as a major advantage of the
BWR over the PWR.

Since a BWR is coupled directly to the turbine of a generating set,
special provision must be made to dispose of steam if the turbogenerator cannot for any reason accept it, or if any malfunction should
occur. A BWR is therefore enclosed – pressure vessel, attached piping
and all – inside a primary containment, which consists of a huge flask-
shaped concrete housing called, confusingly, a 'drywell'. Cavernous
pipes lead from the bottom of the drywell down into a ring-
shaped tunnel, amply large enough to walk through, half-filled with
water. This tunnel is called a 'pressure-suppression pool'. If for any
reason steam or water escapes from the reactor vessel or the pipework,
it is confined in the drywell and channelled down through the pipes
leading into the water in the pressure-suppression pool. Any steam
which gets this far is thereupon condensed, and any excess pressure
it would otherwise exert on the containment is – as the name suggests – 'suppressed'.

The function of the BWR containment is closely associated with
that of the emergency core-cooling systems, provided, like those in a
PWR, to prevent overheating of the reactor core in the event of an
accident. Once again, we shall defer further description of these features until Chapter 6 (pp. 141–3).

Like a PWR, a BWR is refuelled off load, with the reactor shut
down and cooled. Refuelling of a BWR is somewhat more of a chore;
as well as flooding the reactor well, and unbolting and removing the
lid, it is also necessary to lift out and set aside a motley assortment
of steam-processing fittings.

Like the coolant in a PWR, the coolant in a BWR may become
slightly radioactive. Since the primary coolant in a BWR supplies steam
directly to a turbine, some of the radioactivity in the coolant may reach
the turbines. However, in practice most of the radioactivity in BWR
coolant stays in the liquid water, and does not travel with the steam to
the turbine.

The BWR shares with the PWR the drawback of comparatively
low coolant temperature, and resulting inefficiency of conversion of
heat to electricity. A typical BWR output temperature is less than
300°C. On the other hand the BWR also shares with the PWR the
problems associated with relatively high power-density, as we shall
discuss further in Chapter 6 (pp. 141–3). The BWR is also more sus-

ceptible to 'burn-out' or 'steam blanketing', which arises if a layer of steam forms next to the hot fuel cladding. The low heat conductivity of the steam means that the heat is no longer so effectively removed from the fuel, and the fuel temperature may rise suddenly and dangerously.

Design and operation of all types of reactor must take into account the possibility of sudden surges, called transients: temperature transients, pressure transients and so on. This is particularly true for reactors of high power-density, like the light water reactors.

Heavy Water Reactors

CANDU Reactors

The Canadian role in fission research during the Second World War was particularly concerned with heavy water. But after the war Canada decided against embarking on a nuclear weapons programme. Accordingly, with no facilities for uranium enrichment, but with a plentiful supply of indigenous uranium, Canada chose to concentrate on heavy water natural uranium reactors. For some years efforts were directed primarily to fundamental research. However, by the mid 1950s interest began to focus on the development of a power reactor, indeed a family of power reactors, sharing the family name CANDU (for CANadian Deuterium Uranium). The name also serves as a trade-mark, echoing the North American assertion of capability, 'can do'.

The CANDU design came of age in 1971 with the start-up of the first and second of four 508-MWe reactors at Pickering, near Toronto. The Pickering station was then doubled. When the eight reactors are all on stream its total output will be some 4000 MWe. Another CANDU station, Bruce, on Lake Huron, has four 750-MWe CANDU reactors in operation with another four under construction; when they are on stream the Bruce station will probably be the largest single nuclear station in the world. Small CANDU reactors are operating in India and Pakistan and a 630-MWe CANDU at Wolsung in South Korea; another is nearing completion at Embalse in Argentina.

The design used at Pickering and Bruce is called the CANDU-PHW, since it uses pressurized heavy water as coolant (see Figure 7, and the table on pp. 72–3). The heart of the CANDU-

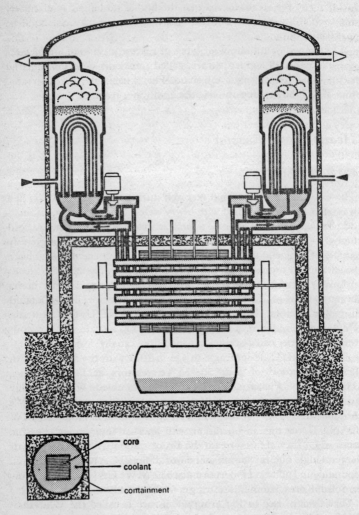

Figure 7   CANDU reactor

PHW is a horizontal cylindrical stainless steel tank, with circular ends. Through this tank, called the 'calandria', run horizontal zircaloy tubes. Inside each of these calandria tubes is another similar tube of slightly smaller diameter; this inner tube is a pressure tube inside which lie twelve short bundles of fuel rods. The fuel rods, natural uranium oxide pellets in zircaloy tubes, form a cylindrical faggot, containing 22 kilograms of uranium oxide. The space in the pressure tube not filled by fuel bundles is taken up by heavy water, flowing through the tube. Emerging from individual pressure tubes at each end of the calandria, the hot heavy water feeds into larger-diameter 'header' pipes which carry it to steam generators.

In a graphite-moderated reactor the core can be made out of solid graphite with holes drilled through it for fuel and coolant. It is not easy to drill permanent holes through heavy water; but comparable geometry is achieved by containing the heavy water moderator in a tank – the calandria – shaped as though it had horizontal holes drilled through it for fuel and coolant.

The moderator circuit is kept cool and at atmospheric pressure, and the space not filled with liquid heavy water is occupied by helium cover gas. Below the reactor core is a dump tank which can accommodate the entire heavy water inventory of the moderator system.

Control rods enter the reactor from above. Only one of the eleven shutdown rods passes between any two calandria tubes; a mechanical distortion of a tube in the event of accident would jam at most two of the eleven rods.

The refuelling system of a CANDU reactor is complex and ingenious. A CANDU is designed to be refuelled continuously on load. The arrangement recalls the earliest plutonium production reactors, although the CANDU technique is much more elaborate and fully automated. At either face of the reactor is a refuelling machine in a shielded vault. One machine rams fresh fuel bundles into one end of the tube, while the other collects used ones as they emerge from the other end. The full machine then feeds the used fuel down a conveyor to eventual storage in a large water-filled cooling pond under the station. In due course it will be necessary to decide what to do with the accumulated used fuel bundles (see pp. 94–100).

Variations of the basic CANDU design included the Whiteshell Reactor WR-1, with an organic fluid coolant, which might lend itself well to a thorium fuel cycle; and the Gentilly-1 CANDU-BLW (for Boiling Light Water), in which a light water coolant was allowed to boil in vertical fuel channels, the steam passing in a direct cycle to a turbogenerator. There is also a British first cousin of the CANDU-BLW, called the Steam Generating Heavy Water Reactor (SGHWR), which uses enriched rather than natural uranium. Only a 100-MWe prototype of the SGHWR exists at Winfrith, Dorset. It started up in 1967, and has a commendable operating history as a working power station. But it seems destined to be the first and last of its line; see p. 165. One other pressure-tube design also deserves mention, a hybrid in which fuel elements lie in vertical pressure tubes filled with light water and surrounded by graphite moderator. The design is called the RBMK, an acronym in Russian. The Soviet Union has built more than a dozen RBMK units of up to 1000MWe, including four in Leningrad.

Fast Breeder Reactors (FBRs)

All the reactors so far described share a common feature. Their physical basis is fission induced by slow 'thermal' neutrons. Such reactors can be called, as a group, 'thermal' reactors. Even in a thermal reactor some of the available neutrons are absorbed by uranium-238, turning it into plutonium-239, which may then fission and make a significant contribution to the total release of energy. But the amount of plutonium created is less than the amount of uranium used up; so such reactors can also be called 'burner' reactors.

We indicated earlier that uranium-238 in this context is a 'fertile' material. In a reactor containing both fissile and fertile material, the comparison between fissile nuclei consumed and fertile nuclei converted to fissile is called the 'conversion ratio'. For instance, if for every 10 uranium-235 nuclei undergoing fission 8 uranium-238 nuclei are converted to plutonium-239, the conversion ratio is 0·8.

In a burner reactor, by definition, the conversion ratio is less than 1. A substantial conversion ratio even when less than 1 is handy. In a CANDU reactor, for instance, before a fuel bundle is discharged, an impressive number of uranium-238 nuclei have already been converted into plutonium-239 and subsequently undergone fission, making a sizeable contribution to the total heat output from the bundle.

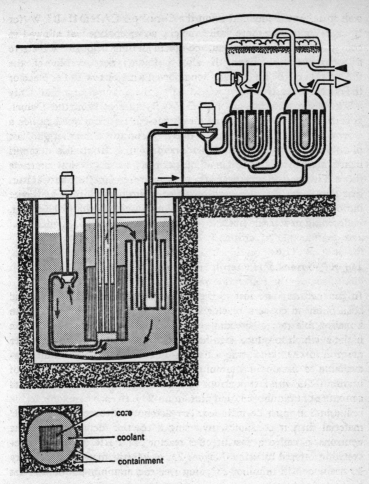

core
coolant
containment

Figure 8   Fast breeder reactor (FBR)

CANDU designers consider this once-through approach a particularly elegant way to utilize plutonium.

It is also possible to design a reactor with a conversion ratio greater than 1: a 'breeder' reactor, which produces more fissile material than it consumes. At the end of its sojourn in the core, fuel from such a reactor emerges containing more fissile nuclei than it contained when new. Of course it also contains the usual complement of ferociously radioactive fission products; recovering the new plutonium is not easy. Nonetheless the concept of the breeder has long played a major role in the plans of the nuclear industry.

The design criteria for a breeder are very different from those which govern the reactor types thus far discussed. As mentioned earlier a thermal neutron is much more likely to rupture a uranium-235 or plutonium-239 nucleus than is a fast neutron fresh from a fission event; hence moderators are used in all burner reactor cores to slow neutrons down. This may suggest that fast neutrons are pretty ineffectual in a chain reaction. But a fission caused by a fast neutron produces on average more new fast neutrons than does a fission caused by a thermal neutron.

Breeding new fissile nuclei in a chain reaction requires, under ideal and unattainable circumstances, exactly two new neutrons from each fission: one to carry on the chain reaction by causing a further fission, and one to transmute a fertile nucleus into a fissile one. (Under such circumstances the conversion ratio is exactly 1 – replacement value.) In fact neutrons are lost to the system by leakage and by 'parasitic absorption' in coolant, reactor structure etc. Accordingly, to achieve a measurable rate of breeding, the reacting system must depend on fissions which produce significantly more than two neutrons per neutron lost while causing a fission. The most obvious combination available is fission of plutonium-239 by fast neutrons. Fission of uranium-235 with fast neutrons is less efficient, but will work. So will a mixture of uranium-235 and plutonium-239. In each case some fertile uranium-238 must be included. A reactor which breeds more fissile material than it consumes, by using a reaction dependent on fast neutrons, is called a 'fast breeder reactor', or FBR. An even better system is offered by using thorium-232 as a fertile material to produce its fissile cousin uranium-233; however the uranium–plutonium fast breeder systems have thus far received the great majority of attention and development.

As it happens, the first reactor ever to power electric generating equipment was a fast breeder reactor. On 20 December 1951, at the National Reactor Testing Station (NRTS) in Idaho, the Experimenta' Breeder Reactor-1 (EBR-1) produced enough electricity to light fou

25-watt bulbs. The first true power reactors based on the fast breeder principle were the British Dounreay Fast Reactor in Caithness on the north Scottish coast, the Experimental Breeder Reactor-2 (EBR-2) at NRTS, Idaho, and the Detroit Edison Enrico Fermi-1 reactor near Detroit, Michigan. The Dounreay Fast Reactor started up in 1959 and the EBR-2 in 1963. However, the Detroit Edison reactor which was intended to be the prototype of a full-fledged commercial fast breeder reactor, experienced endless trouble including an accident that might have necessitated the evacuation of Detroit (see pp. 135-7). It has since been permanently shut down.

A new generation of prototype fast breeder power reactors emerged in the mid 1970s, including the Soviet BN-350 reactor at Shevchenko on the Caspian Sea, the French 250-MWe Phénix reactor at Marcoule, and the British 250-MWe Prototype Fast Reactor (PFR) at Dounreay, all of which went critical between November 1972 and February 1974. The Fast Flux Test Facility (FFTF) at Hanford in the USA started up at last in 1981 after many years' delay; but the status of the prototype power station long planned for Clinch River in Tennessee remains problematical. Even larger fast breeder power stations are the BN-600 in the Soviet Union, which started up in 1980, and the 1200-MWe Super-Phénix at Creys-Malville, France, due to start up in 1984.

The fundamental difficulty facing the designer of a fast breeder reactor is that it takes 400 times as many fast neutrons as thermal neutrons to cause one fission. Accordingly, a much higher neutron density must be created. Furthermore, newly emergent neutrons must avoid collisions which would slow them down before they strike other fissile nuclei. The core of a fast breeder reactor must thus be far more compact than that of any power reactor thus far described. Not only does it contain no moderator: it also contains a minimum of other structural material, and as little coolant as suffices to carry away a fiercely intense output of heat. The technological challenge is one of the most demanding ever encountered.

The overall layout of a fast breeder is a compact core of concentrated fissile material surrounded by a 'blanket' of fertile material to catch the neutrons pouring from the core (see Figure 8, and the table on pp. 72–3). The design currently most favoured, of which all the above-named FBRs are examples, uses molten metal (usually sodium) as coolant.

(The Dounreay Fast Reactor used an alloy of sodium and potassium – usually called 'nak' for Na and K, the chemical symbols of the metals – which is liquid at room temperature.) Such a 'liquid metal fast breeder reactor' (LMFBR) is to be sure not the only possible design; a gas-cooled FBR is also possible and has its advocates. But liquid metal coolant has obvious advantages. Liquid sodium, being a metal, has a high thermal conductivity; even without moving through an FBR core it can siphon out considerable heat. Furthermore, since it boils at the high temperature of 990°C it need not be pressurized, which considerably reduces one major engineering problem.

On the other hand, sodium does have drawbacks. As every boyhood chemist knows, sodium reacts enthusiastically with water; it reacts likewise with a wide range of other materials. Accordingly, although the sodium coolant is not itself pressurized, its open surfaces in an LMFBR circuit are covered by an inert gas such as argon – which in turn tends to get swept into the flowing sodium and cause unwanted bubbles. Unlike gases or water (light or heavy), sodium is opaque, making remote inspection of reactor internals peculiarly difficult. Sodium coolant must not, of course, be allowed to cool below its melting point of 97·5°C anywhere in the circuits, or it solidifies.

Sodium does not readily absorb fast neutrons – if it did it could not be used in a fast-neutron core – but when it does it becomes sodium-24, which is an intensely radioactive gamma emitter. As a result the primary sodium coolant must be confined entirely within the biological shielding of the core. This necessitates a second sodium circuit, with a heat exchanger inside the biological shielding – but itself shielded from neutrons – to pick up the heat from the radioactive primary sodium and carry it out through the shielding to a second heat exchanger in which steam is generated. Steam generators in which molten sodium and water are separated only by thin tube walls must be fabricated to very high standards; steam generators have proved to be one of the most troublesome features of LMFBRs.

The world's longest-serving FBR was the UK Dounreay Fast Reactor, whose experimental programme was concluded in 1981 with its final shutdown. Its core was only 53 centimetres high, hexagonal, 52 centimetres across each face; a person could easily put his arms around it – although he would have been ill-advised to do so. Its maximum power output was 60 MWt, giving 14 MWe. That 60 MWt was generated, note, in a core whose volume was only about 110 litres –

a power density of over 500 kilowatts per litre, well over one hundred times that in a Magnox core. But the Dounreay Fast Reactor, while supplying a nominal power output, was primarily intended as a laboratory for development of FBR fuel and other technology. It was clear from an early stage that uranium metal fuel, as in the Dounreay Fast Reactor, would not permit operating temperatures high enough to achieve the desired electrical efficiency. Accordingly, the 250-MWe Prototype Fast Reactor (PFR) uses oxide fuel, with its higher melting point. For the PFR the fuel is not just uranium oxide, but a mixture of the oxides of uranium and plutonium. The uranium is only natural uranium; 'depleted' uranium is even better, mixed with enough plutonium to provide the fissile material required. 'Depleted' uranium is uranium from which some of the uranium-235 has been removed to be incorporated in enriched uranium (see pp. 78–82). The low thermal conductivity of the oxide mixture necessitates making the individual stainless steel fuel pins very narrow to keep the interior temperature from reaching embarrassing heights; a PFR fuel pin is less than 6 millimetres in diameter. There are 4·1 tonnes of mixed oxide fuel in the core, including the equivalent of 1·1 tonnes of plutonium-239 oxide.

The core-blanket array is enclosed in an open-topped tank full of molten sodium, and sits in a much larger pot of molten sodium. The sodium emerges from the top of the fuel and blanket assemblies and flows through intermediate heat exchangers, giving up its heat to secondary, non-radioactive sodium. Three primary sodium pumps stir the primary potful. The secondary circuits carry the heat out through the shielding to the steam generators. No pipes or other penetrations enter the primary pot below the level of the sodium, minimizing the possibility of any loss of primary coolant. Above the core, in the reactor roof, is a 'rotating shield' from the bottom of which projects the re-fuelling assembly.

The PFR, like the other LMFBRs of its generation, is at least partially an experimental facility, aimed at establishing the criteria for a commercial fast reactor. One of these criteria is the rate of breeding achievable. A common measure of this important characteristic of performance is the so-called 'doubling time': the time taken for a breeder reactor to double the amount of fissile material associated with its operation. This inventory of fissile material includes that within the reactor core at a given time, in irradiated fuel elements in the cooling pond, in transit

Table: Typical reactors

Type	Magnox Reactor	Advanced Gas-cooled Reactor (AGR)	Pressurized Water Reactor (PWR)
Name of example	Dungeness A (UK)	Hinkley Point B (UK)	Zion 1 (USA)
Heat output	840 MWt	1494 MWt	3250 MWt
Electrical output	275 MWe	621 MWe	1050 MWe
Efficiency	32.7%	41.6%	32.3%
Fuel	Natural uranium metal rods clad in 'Magnox' alloy	Uranium oxide, 2% enriched, clad in stainless steel	Uranium oxide, 3% enriched, clad in zirconium
Weight of fuel	304 tonnes	113.7 tonnes	99 tonnes
Fuel burn-up	3850 megawatt-days per tonne	18 000 megawatt-days per tonne	21 800 megawatt-days per tonne
Moderator	Graphite	Graphite	Water ('light' water)
Core dimensions	13.8 metres diameter 7.4 metres high	11 metres diameter 9.8 metres high	3.35 metres diameter 3.6 metres high
Peak power density	1.1 kilowatts per litre	4.5 kilowatts per litre	102 kilowatts per litre
Coolant	Carbon dioxide gas	Carbon dioxide gas	Water ('light' water)
Coolant pressure	19 atmospheres	40 atmospheres	150 atmospheres
Coolant outlet temperature	245°C	634°C	318°C
Vessel	Welded steel, 0.102 metres thick	Pre-stressed concrete, 5 metres thick	Welded steel, 0.203 metres thick
Refuelling	On load	On load	Off load
Comments	Low power density and mass of graphite mean slow temperature rise in fault conditions. Main hazard is low melting point and ignition temperature of Magnox, if air should enter breach in cooling circuit.	Low power density and mass of graphite mean slow temperature rise in fault conditions. Entire primary cooling circuit is enclosed in vessel, and oxide fuel in stainless steel has wide safety margin above operating temperature before melting temperature.	Very high power density. Loss of coolant pressure also involves loss of moderator – shuts down fission reaction but loses heat sink. Fault conditions may produce very rapid temperature rise, possibly even to melting temperature of oxide fuel. Heavy section welded steel pressure vessel requires very high-quality construction, because of very high operating pressure.

Boiling Water Reactor (BWR)	CANDU Reactor	Liquid Metal Fast Breeder Reactor (LMFBR)
Browns Ferry 1 (USA)	Pickering 1 (Canada)	Phénix (France)
3293 MWt 1065 MWe	1744 MWt 508 MWe	563 MWt 233 MWe
32.3% Uranium oxide, 2.2% enriched, clad in zircaloy	29.4% Natural uranium oxide clad in zircaloy	41.4% Mixed uranium and plutonium oxides, 20–27% effective enrichment, clad in stainless steel
169 tonnes 19 000 megawatt-days per tonne Water ('light' water) 4.8 metres diameter 3.7 metres high 49 kilowatts per litre	92.6 tonnes 7000 megawatt-days per tonne Heavy water 6.4 metres diameter 5.9 metres long 16.2 kilowatts per litre	4.3 tonnes 100 000 megawatt-days per tonne None 1.4 metres diameter 0.85 metres high 646 kilowatts per litre
Water ('light' water) 68 atmospheres	Heavy water 85 atmospheres	Liquid sodium 1 atmosphere
285°C	293°C	562°C
Welded steel 0.159 metres thick	Zircaloy pressure tubes 0.1 metres in diameter, 5 millimetres thick	Cylindrical stainless steel pot 12 metres in diameter, 12 metres high
Off load High power density. Loss of coolant pressure also involves loss of moderator – shuts down fission reaction but loses heat sink. Fault conditions may produce rapid rise in temperature, possibly even to melting temperature of oxide fuel. Heavy section welded steel vessel requires very high-quality construction because of high operating pressure.	On load Rather low power density, and cool moderator in separate system, mean slow temperature rise in fault conditions. Pressure tube construction means less likelihood of propagation of a flaw from one tube to others. Fabrication of pressure system involves simpler configurations than those of full-size pressure vessel, despite high coolant pressure.	Off load Power density 10–100 times that of 'thermal' reactor designs, but metallic heat conductivity of sodium provides cooling even if circulation fails. System at atmospheric pressure, so no depressurization problem. Fuel is concentrated fissile material – unlike that of 'thermal' reactors. Change of geometry – if coolant flow is interrupted – may produce increase in rate of fission reaction, perhaps very rapid increase. Fuel also presents safeguards problem because of possible misuse of plutonium.

to the reprocessing facility, within the reprocessing facility, in transit to the fuel fabrication plant, within the fuel fabrication plant, in transit back to the reactor and awaiting insertion into the core. The total aggregation of fissile material outside the core is the 'pipeline inventory' associated with the FBR. As a rule, in addition to the fissile material within the reactor, there will also be three or four times this amount outside it at other stages of the fuel cycle – perhaps, for a reactor of the size of the PFR, four to five tonnes of plutonium.

The breeding gain is the additional proportion of plutonium contributed during the time a fuel charge spends in the FBR. The smaller this breeding gain, the greater the number of cycles required to double the total amount of plutonium. Accordingly, there are two avenues available through which the doubling time can be shortened: the breeding gain can be increased, or the length of a given fuel cycle can be shortened. Increasing the breeding gain within the reactor means, essentially, operating at a higher neutron-flux; this means decreasing the space between fuel pins, while at the same time requiring a much increased rate of heat removal, criteria clearly in conflict. The only parts of the fuel cycle which can be shortened are those which occur outside the reactor. The obvious step to shorten is the sojourn in the cooling pond. Unfortunately, shortening this sojourn – to as little as thirty days, as has been suggested – means transporting irradiated fuel which is still very radioactive indeed, with all that that entails.

Doubling times in the present generation of FBR cores seem unlikely to be much less than twenty years; some would put the figure much higher. FBR designers are aiming for doubling times of less than ten years; but the engineering required, and the implications for safety of working to such close tolerances, may prove daunting.

3 · The Nuclear Fuel Cycle

Undoubtedly the most extraordinary things that happen to reactor fuel happen within the core of an operating reactor. But a great deal happens to it outside the reactor, both before and after its sojourn in the core. The odyssey of the fuel material, from its origin in the earth's crust, takes it from a mine, to a mill, possibly through a specialized facility called an enrichment plant, and through a fuel fabrication plant before it enters the reactor. When it emerges from the reactor it may go into storage or into another specialized facility called a reprocessing plant. Some of the material thereafter reaches a theoretically final resting place, while the rest may re-enter the process at an earlier stage. The whole succession of processes, with the transport which links them, is called the nuclear fuel cycle. In practice it is not very cyclic; but the possibility exists of making it much more so, provided certain problems – both technical and otherwise – can be overcome. Present policies within the nuclear industry are generally directed towards this end. But, cyclic or otherwise, the nuclear fuel cycle outside the reactor gives rise to many of the most controversial aspects of nuclear technology. In the following pages we discuss the fuel cycle, and some of the problems which arise in it.

Uranium Production

Uranium is found in nature as mineralization in sandstones, in quartz pebble conglomerate rocks, and in veins, and to a smaller extent in other types of deposit. There are significant uranium reserves in the USA, Canada, southern Africa, Australia, France and elsewhere. High-grade uranium ores contain up to 4 per cent uranium; but known reserves of this quality have been largely worked out, and ore grades ten times lower, 0·4 per cent and less, are now being worked. Still lower grades – down to 0·01 per cent and less – are also being noted for development.

Uranium ore deposits are found by a variety of exploratory techniques. The folklore image of the uranium prospector with his Geiger counter, picking his way over the hillside listening for clicks, has little to do with contemporary uranium prospecting. Uranium exploration usually begins in the air, looking for abnormal traces of airborne radio-activity given off by the decay products – so-called 'daughters' – of uranium. Airborne instruments look for tell-tale gamma rays and other

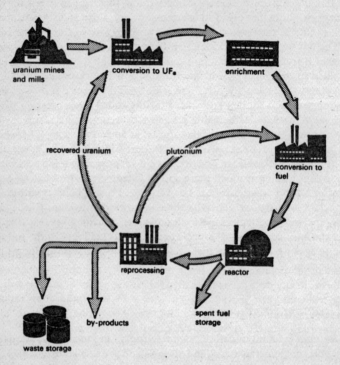

Figure 9   The nuclear fuel cycle

evidence of radioactivity. More evidence is assembled on the ground, by studying likely geological formations, by testing samples chemically, and ultimately by drilling.

Uranium ore is extracted by surface or underground mining. The crude ore is fed into a series of crushing mills, which grind it to the consistency of fine sand. Chemical solvents then dissolve out the uranium, which emerges from the process in the form of a mixture of uranium oxides with a chemical formula equivalent to U_3O_8. This oxide mixture, usually called 'yellowcake', forms the raw material for all the succeeding processes that lead eventually to the reactor core and the chain reaction. Yellowcake contains 85 per cent uranium by weight. Besides the yellowcake, there remains after extraction some one hundred times its weight of residual sand, called 'tailings' – which also contains the radium which had accompanied the uranium. There also remains, per tonne of ore, over 3700 litres of liquid waste, which is both chemically toxic and radioactive. A uranium mine and associated mill may produce over 1000 tonnes of uranium per year, from at least 250 000 tonnes of ore.

Hazards arise at several stages in the uranium production process. The first of these arises from the uranium ore itself, *in situ* and subsequently. When uranium-238 undergoes alpha decay, it produces a succession of further alpha-emitters, including radium-226 and its immediate daughter-product, the chemically inert but radioactive gas radon-222. Any aggregation of uranium which has remained for some time undisturbed – such as a geological deposit – therefore exudes this radioactive gas. When a uranium ore deposit is broken up in mining the escape of the radon is facilitated. Radon-222 is an alpha emitter with a half-life of less than four days, which produces its own radioactive 'daughters'. These radon daughters are however solids. When a radon-222 nucleus in the air emits an alpha particle, the resulting nucleus of polonium-218, being momentarily electrically charged, adheres to any dust particle nearby. Accordingly, air containing radon also contains dust particles laden with intensely radioactive radon daughters. Underground uranium miners who are permitted to inhale such air have proved appallingly susceptible to lung cancer.

The first evidence of this effect was established by 1930, after medical investigations of miners working deposits in Joachimsthal, in Germany. A similar effect has since appeared in the miners working deposits in the south-western USA after the Second World War. Inadequate ventilation and insufficient expenditure on mine safety were blamed for the lung cancer deaths of over 100 American uranium miners; out of a total of some 6000 men who had worked in American underground uranium mines in the boom years of the 1950s the US Public Health Service estimated at the end of the 1960s that from 600 to 1100 would die of lung cancer because of radiation exposure on the job. In Canada the

Royal Commission on the Health and Safety of Workers in Mines – the Ham Commission after its head – in 1976 devoted an entire chapter of its long report to 'Lung Cancer and Ionising Radiation in the Uranium Mines', making twenty-three substantive recommendations about desirable improvements.

Uranium mine tailings also present a problem. The military rush for uranium in the USA led to accumulation of vast piles of tailings; estimates range as high as 90 million tonnes, much of this piled on river-banks in the south-western USA. The consequent radioactive pollution of waterways has represented a serious problem; at one stage inhabitants downstream in the Colorado river basin were exposed, through their drinking water, to three times the ICRP (see Appendix B, p. 232) maximum permissible intake of radium – which is a bone-seeking radio-nuclide even more dangerous than strontium-90. Canadian tailings piles likewise caused concern; indeed, the Ontario Royal Commission on Electric Power Planning – the Porter Commission – observed in 1978 that uranium mine tailings might be much the most serious long-term radioactive waste problem. Dry tailings are still blown freely by the wind across many inhabited areas of the south-western USA. The tail-ings piles will remain dangerously radioactive for tens of thousands of years.

Meanwhile it was discovered in the 1960s that the sandy tailings had been used as fill beneath the foundations of many buildings in many communities, notably Grand Junction, Colorado: buildings including homes, schools and hospitals. The radon gas emanating from these building structures into the air indoors now exposes the local inhabit-ants – including children – to exactly the same radon daughter-products which have already been responsible for thousands of lung-cancer deaths of miners from Joachimsthal to Grand Junction. Some govern-ment aid has been available to reconstruct these radioactive buildings; but many residents would prefer that nothing be said about it, because of the effect on property values.

Uranium Enrichment

As indicated in Chapter 1, the fissile uranium-235 nuclei in natural uranium – 7 out of 1000 nuclei – are too dilute to support a chain reac-tion. Their effectiveness can be increased by interspersing the uranium fuel with a moderator to improve the neutron economy, as already described (pp. 32–3). Alternatively, or in addition, it is possible to increase the proportion of uranium-235 nuclei in the material. This process is called 'uranium enrichment'. Indeed, for weapons applications, it is possible, and often necessary, to have uranium which consists almost wholly of the 235 isotope; such weapons applications use uranium which is at least 90 per cent uranium-235.

Bringing about this increase in the concentration of the 235 isotope is not, however, easy. Chemically it is very difficult; in chemistry uranium-235 and uranium-238 are virtually identical. Only their minute difference in mass – 3 units in 235 – can be used as a basis for separation. There are several physical phenomena in which this minute mass difference produces a measurable difference in behaviour between the two isotopes. Of these the phenomenon of earliest large-scale interest was the rate of diffusion through a thin porous membrane. The lighter uranium-235 diffuses just slightly more swiftly through such a membrane; this effect is the basis for what are arguably the largest industrial establishments in the world, the 'gaseous diffusion plants'. There are three such plants in the USA; one in the UK, at Capenhurst in Cheshire, now shut down; two in France, at Pierrelatte and Tricastin; two in the Soviet Union; and at least one in China.

The details of gaseous diffusion technology are – because of their military implications – still to a considerable extent secret. The basis of a gaseous diffusion plant is very simple: a metal-walled cell, with a thin membrane of porous metal dividing it in two. (Fabrication of such membranes, which must withstand lateral pressures and chemical corrosion while providing a selective diffusion barrier, is one area where much detail is still not publicly available.) In order to utilize the different diffusion rates of the two uranium isotopes, it is necessary to convert the original yellowcake, solid uranium oxides, into uranium hexafluoride, UF_6. This compound, called 'hex' for short, is the simplest compound of uranium which can be easily vaporized. Furthermore, fluorine has only one stable isotope; so the different diffusion rates of hex molecules will depend only on the difference between the uranium isotopes involved. It must be added that hex is a viciously corrosive, reactive gas, requiring very careful handling and high-quality metallurgy in the vessels through which it travels.

Under controlled pressure hex enters one chamber of a diffusion cell. It diffuses through the membrane into the other chamber, the lighter

235 isotopes diffusing slightly faster than the heavier 238 isotopes. In a given cell the concentration of 235 can be increased, however, only by about one part in a thousand. Accordingly, the diffusion process must be repeated thousands of times. A cascade arrangement is set up. Gas from the high-pressure chamber of a cell, slightly depleted of the 235 isotope, is piped back to earlier cells; gas from the low-pressure chamber, slightly enriched in the 235 isotope, is piped onwards to later cells. By this means, using thousands of pumps and condensers, it is possible to raise the proportion of 235 isotope to more than 99 per cent. Since pumping heats the gaseous hex the plant must also include large-scale cooling systems.

The uranium whose share of 235 nuclei has been reduced is called, as mentioned earlier, depleted uranium. One factor affecting the performance of a gaseous diffusion plant is the 'tails assay' – the level at which the percentage of 235 is so low that it is no longer worth trying to extract any more 235 from the hex. This tails assay is usually somewhere between 0·2 and 0·3 per cent 235, compared with 0·7 per cent in natural uranium. If the depleted hex is discharged from the plant when it still contains 0·3 per cent 235, more yellowcake will be required to produce a given amount of uranium enriched to a given level; on the other hand if the depleted hex is not discharged until its tails assay is down to 0·2 per cent 235, part of the plant operates with very depleted hex, from which it is even more difficult to extract a useful amount of 235.

In the early stages of enrichment, diffusion cells must be comparatively large; the desirable uranium-235 nuclei are accompanied by a comparatively cumbersome cloud of uranium-238 fellow-travellers. As the proportion of 235 increases, the total mass of hex which must pass through successive cells decreases; the high-enrichment end of the plant uses comparatively small cells, in which only the 235 nuclei remain, with a few stragglers of 238. For this reason the early stages of enrichment, to 3 or 4 per cent uranium-235, require as much pumping as all the stages from this level onwards. The effort expended in the enrichment process is measured in units of 'separative work'; the throughput capacity of the plant is measured in units of separative work per year. Separative work is loosely correlated with the total energy required to carry out an operation – energy to run pumps, etc. In general a comparatively large amount of hex enriched to a few per cent requires the same amount of separative work as a comparatively small amount of hex enriched to 90-plus per cent.

All the first generation of gaseous diffusion plants were built under military auspices. Their electrical requirements are awesome; the Oak Ridge plant, in full operation, requires some 2000 megawatts of electricity – enough to power a sizeable city. (Electricity for the Oak Ridge plant is largely provided by fossil-fuel power plants burning strip-mined coal, a nicely ironic touch.) A gaseous diffusion plant likewise takes up an impressive area, as much as half a square kilometre. However, because of the differences between the low-enrichment end of the plant and the high-enrichment end, such plants are not easy to convert from production of strictly military weapons-material, involving enrichment to more than 90 per cent uranium-235, to production of fuel for power reactors. Water reactors and AGRs require for their fuel a level of enrichment of only 2 to 4 per cent. In consequence, although the first-generation gaseous diffusion plants are in theory available to service present power reactors, other approaches to enrichment are now attracting attention.

France, in partnership with several other countries under the corporate name of Eurodif, built a large diffusion plant at Tricastin, designed especially for production of reactor fuel and different in detail from the military plant at Pierrelatte. France also planned a further diffusion plant to sell enrichment services to foreign customers; but this consortium, called Coredif, came a cropper with the fall of the Shah of Iran, whose government had put up 20 per cent of the investment capital for the plant. The consequent financial dispute remains unresolved.

Meanwhile an alternative enrichment technology is making its first contribution to the present-day nuclear fuel cycle. Like the gaseous diffusion process this alternative requires thousands of stages in cascade; the stages this time consist of gas centrifuges. When uranium hexafluoride gas enters a spinning centrifuge, the uranium-238 hex molecules tend to drift to the outer perimeter of the centrifuge chamber, leaving the lighter uranium-235 hex molecules closer to the axis of the chamber. Piping channels the axial hex, slightly enriched, onwards to successive centrifuges, and the perimetral hex, slightly depleted, backwards – just as in the cascades of a diffusion plant. It is claimed that the centrifuge method consumes only one tenth of the energy required for diffusion, a major advantage for the centrifuge approach. The Almelo Treaty in 1970, signed by the UK, the Federal Republic of Germany and the Netherlands, created two tripartite consortia, each country holding a one-third partnership: URENCO, to sell enrichment services, and CENTEC, to manufacture the hundreds of thousands of centrifuges required. URENCO centrifuge enrichment plants at Capenhurst in Cheshire and at Almelo in the Netherlands delivered their first separative work in the late 1970s. Both plants continue to expand, as new contracts require. A third plant is planned for Gronau in the Federal Republic.

Other techniques are also under development. One is based on deflection of gas sprayed from a nozzle: the lighter hex-235 molecules are more easily deflected. A small prototype nozzle plant has been built in the Federal Republic, and a larger plant employing the same principle is in operation under conditions of great secrecy at Valindaba in South Africa. Undoubtedly the most exotic technique for isotope separation is based on lasers. A laser can be tuned so finely that its radiation ionizes uranium-235 atoms or hex molecules while not ionizing uranium-238 atoms or hex. It is then necessary somehow to use the handle provided by the electric charge on the ionized atoms or molecules to sift the 235 atoms or molecules out of the cloud. Major research programmes are known to be under way at the US nuclear weapons laboratories at Los Alamos and Livermore, at Harwell in the UK, and possibly in Israel and elsewhere. The curtain of secrecy around laser enrichment is akin to that surrounding the Manhattan Project itself; but oblique reports indicate that the technology is proving entirely feasible. The secrecy is all too understandable: unlike the other technologies mentioned, a single stage of laser enrichment could conceivably bring about almost complete separation of fissile uranium-235 and non-fissile uranium-238, offering an alarmingly short cut to weapons-material, even from uranium ore.

Heavy Water Production

Uranium isotopes are not the only ones requiring separation for nuclear applications. At the other end of the table of elements come the isotopes of hydrogen – of which the second, deuterium, is the best neutron moderator of all. The American and Canadian reactor designs offer in this context a tidy contrast: whereas the Americans enrich the fuel and take the moderator as it comes, the Canadians take the uranium as it comes and, so to speak, enrich the moderator.

The difference in mass between an ordinary hydrogen nucleus and a nucleus of heavy hydrogen or deuterium is proportionally very large; a deuterium nucleus is about twice as massive as a nucleus of ordinary hydrogen. As a result certain types of chemical interchange can be used to separate the light and heavy hydrogen nuclei. The Girdler-Sulphide (GS) process now in large-scale use employs the two chemically similar molecules water and hydrogen sulphide. The former consists of two hydrogen atoms bonded to an oxygen atom, the latter of two hydrogen atoms similarly bonded to a sulphur atom. In a mixture of water molecules and hydrogen sulphide molecules, the distribution of the hydrogen isotopes between the oxygen and sulphur atoms depends on the temperature. At low temperatures – about 25°C – the liquid water contains proportionally more deuterium than it does at higher temperatures – about 100°C. This shift of equilibrium can be used to transfer deuterium atoms out of one batch of water and into another, using hydrogen sulphide as a sort of conveyor belt.

First the water and hydrogen sulphide are mixed together at the lower temperature; deuterium shifts from hydrogen sulphide into water. Some of the enriched water is led off for further enrichment. The rest is fed into a tower at the higher temperature; deuterium now shifts from this water into the hydrogen sulphide. This enriched hydrogen sulphide in turn shuttles back to enrich more water. The depleted water can be discarded, and the enriched water fed onward through a cascade, successively boosting its percentage of deuterium.

Whereas the enrichment of uranium becomes easier the higher it gets, as far as mass transport is concerned, the enrichment of water gradually gets too cumbersome, as the water–hydrogen sulphide exchange reaction becomes inefficient. By this time it is however possible to carry out fractional distillation, utilizing the significantly higher boiling temperature of deuterium oxide – about 101·4°C – to boil away much of the remaining ordinary water. Electrolysis can refine this to a final composition of 99·75 per cent deuterium oxide; by this stage electrolysis is a comparatively inexpensive and efficient way to dispose of the remaining ordinary hydrogen.

There are perhaps a dozen heavy water production plants in all – in the USA, Canada, France, India and elsewhere. The continuing international interest in heavy water reactors – and the high efficiency of such reactors for plutonium production – seem likely to continue to keep current production capacity, of the order of 300 tonnes of deuterium oxide per year per plant, fully occupied. On the other hand, heavy water is intended to be a permanent part of a reactor system; unlike enriched fuel it is not 'consumed'. Once a reactor is equipped with its operating complement of heavy water its only requirement from then on is enough to replace losses in refuelling and inevitable leakage. Since heavy water now costs upwards of £100 per kilogram operators strive to minimize such losses.

Fuel Fabrication

Fabrication of fuel for reactors is now a major – and complex – industrial process in its own right. In Chapter 2 (pp. 52–3) we encountered some of the determinants affecting reactor fuel and its cladding: ease of heat removal, durability when subject to radiation damage, chemical stability, and physical and mechanical properties which lend themselves to economical fabrication. An additional requirement, at every stage, is establishing and maintaining high purity in the materials, to keep them free of neutron-absorbing impurities. Fuel fabrication facilities accordingly strive to carry out the relevant industrial processes in conditions of cleanliness like those of an operating theatre.

Among the present range of power reactors the only large ones using uranium metal fuel are the UK Magnox reactors, and their French cousins. The awkward metallurgy of uranium has already been mentioned. Nevertheless uranium can be fabricated by common metalworking techniques.

The uranium fuel of water-cooled reactors – PWRs, BWRs, and CANDUs – is in the form of uranium dioxide. Uranium dioxide powder is made from uranyl nitrate solution, which may originate either in a uranium mill (from natural uranium), an enrichment plant (from enriched uranium hexafluoride), or a reprocessing plant (see pp. 86–9). Fabrication by powder techniques is employed to form the desired shapes – for instance the short cylindrical pellets described in Chapter 2. Baking at high temperature produces stable, dense pellets – the denser the better; high density facilitates the chain reaction, improves the generally poorer thermal conductivity, and also helps to retain the gaseous fission products which accumulate in the fuel material.

If uranium metallurgy is awkward, plutonium metallurgy is positively fiendish. The metal occurs in six different crystal phases, whose properties change drastically with temperature; two phases even contract, rather than expanding, as temperature increases. Its thermal conductivity is low, its melting point is low, it oxidizes violently on contact with air, and – when its peculiarly vicious radiotoxicity is added to the mix – all in all it is a material without many redeeming virtues. Of course, in the one-shot chain reaction of a fission bomb most of these problems are overcome in a flash. But for the controlled chain reaction in a reactor the choice falls not on the metal but on the dioxide.

Its fabrication is a much more demanding process than that of uranium, requiring much tighter control of quantities and ancillaries. Precautions must prevent not only the escape of the toxic material but also inadvertent juxtaposition in undesirable geometries; unlike most power reactor uranium, power reactor plutonium is mostly fissile nuclei, and may easily come together in quantity in such a way as to achieve criticality. The consequent barrage of neutrons and gamma rays could cause serious injury or death to anyone near by. This is particularly hazardous in the case of aqueous solutions of plutonium compounds, since the water acts as a moderator. However, with appropriate precautions plutonium dioxide can be processed like uranium dioxide; indeed, the two oxides, mixed in suitable proportions, are an effective fuel material.

Fuel rods or fuel pellets, once fabricated, are clad as described in Chapter 2 (pp. 52 – 3), and where appropriate arranged in assemblies for transport to the reactor.

Transport

One of the major advantages claimed for central electricity generation with a nuclear heat source is the relatively small bulk and mass of fuel and waste that must be transported to and from the station. A fossil-fuel station requires so much coal or oil that it is economically advisable to situate the station near the fuel supply. A nuclear power station, on the other hand, requires at most one or two shipments a week; furthermore the shipments away from the station are far more massive and bulky than those to the station. Fresh fuel elements, only minimally radioactive, can be and are shipped in ordinary cases like any other cargo. But once irradiated they must be heavily shielded, so that a shipment of two tonnes of irradiated fuel requires a fifty-tonne steel shipping cask.

Fresh reactor fuel and fuel materials are shipped by rail, by road, by water and by air, in increasing quantities every year, between various parts of the fuel cycle. Apart from the usual protections against low-level radioactivity around the shipping cases, the main technical consideration is to guard against the stacking of cases so close together that the aggregation of fissile material can reach criticality (but see pp. 197–201). Elaborate codes of practice are published to provide appropriate technical guidelines.

The shipping of irradiated fuel is something else again. The irradiated fuel must be handled remotely at every stage. For short journeys irradiated fuel elements usually travel in massive water-filled casks which are vaned on the exterior to help dissipate the decay heat. For longer journeys, especially those by sea, the casks must be coupled to cooling circuitry. Since an accident involving a shipment of irradiated fuel could release dangerous amounts of radioactivity, shipping casks must pass severe tests, typically a thirty-minute fire after a ten-metre fall.

Spent Fuel

One feature distinguishes nuclear power technology from all others: the left-overs. Unlike the ash, say, from a coal-fired power station, the used fuel from a nuclear power station contains both potentially valuable material and uniquely troublesome waste. Recall that the first large reactors were built expressly so that, under neutron bombardment, the uranium-238 in the fuel would be transmuted into plutonium-239. This plutonium had to be recovered from the fuel. When the first power reactors came into operation in the mid 1950s, uranium was still in short supply; so it made sense to recover also the unused uranium-235 which was still left in the fuel, after poisoning of the chain reaction by fission products and other effects had made it necessary to remove the fuel from the reactor. It thus became customary to assume that spent fuel from power reactors, like that from plutonium production reactors, should be 'reprocessed'.

A nuclear fuel reprocessing plant is a chemical plant – but no ordinary chemical plant. Because its raw material, irradiated reactor fuel, is intensely radioactive, all the operations must be carried out by remote control, behind heavy shielding. The process equipment must be highly reliable, and require a minimum of maintenance. Once in operation it is contaminated by the radioactivity, and any malfunction necessitates months, or indeed years, of decontamination before it can be set right. Accordingly, the process line uses as few mechanical parts as possible, and depends instead on gravity flow and simple valves.

Different designs of fuel require different handling. The British reprocessing plant at Windscale was originally set up to process metal fuel elements from plutonium production and Magnox reactors. Magnox fuel is stored in a cooling pond adjacent to the reprocessing plant. When ready for reprocessing it is transferred under water, by operators watching on closed-circuit television, into the building and up into the first of a series of 'caves' or 'hot cells'. The walls of the caves are of concrete some two metres thick, to intercept the gamma radiation from the fission products in the fuel. Once in the caves the fuel can be observed through special windows built into the cave walls. Each window is like a large aquarium, filled with a solution of a chemical such as zinc bromide, which is virtually transparent to visible light but strongly absorbs the very short wavelengths of gamma radiation.

A Magnox fuel element entering the reprocessing caves is picked up by remote control and dropped on a stripping machine which chops off the ends of the element and unzips the Magnox cladding as easily as peeling a banana. The contaminated cladding drops on to a conveyor belt to be transported to another building near by, which looks like an aircraft hangar but is actually a heavy concrete storage bin. The bare fuel rod is loaded into a transfer magazine and then dropped into a vat of nitric acid, which dissolves it ready for reprocessing. The nitric acid is mixed with a solution of an organic solvent; in the Windscale plant this solvent has a polysyllabic name which is unceremoniously abbreviated to TBP/OK. The uranium and plutonium cross over into the TBP/OK, leaving behind about 99·96 per cent of the fission products in the water-based acid. This acid stream carries these fission products out of the reprocessing plant; its subsequent progress will be described in the following section.

Almost all the uranium and plutonium (not quite all – see p. 98) are now in the TBP/OK stream, flowing under gravity from one section of the plant to the next. After another pass through a similar 'solvent extraction' stage, to remove lingering fission products, this stream now encounters another water-based solution; the plutonium and uranium at this point part company, the plutonium returning to the water solution leaving the uranium behind in the TBP/OK. Various chemical manoeuvres occur en route, to bring about these shifts of loyalty. Eventually, after repetitions of some steps, the uranium emerges in one stream and the plutonium in another, as uranyl nitrate solution and plutonium nitrate solution respectively. The uranyl nitrate is converted back into solid uranium oxide and stored for possible future use; the plutonium nitrate may be likewise converted back to solid oxide, or may be kept in nitrate solution, ready to be made into fast reactor fuel (or weapons). Process streams of concentrated fissile material – especially plutonium – must be designed to guard against accidental criticality. Reprocessing plants have elaborate alarm systems to warn personnel in the event of a criticality accident, which may of course be quite invisible, despite the fusillade of neutrons and gamma rays.

Reprocessing oxide fuel is more difficult than reprocessing Magnox fuel, both because of the design of the fuel and because spent oxide fuel, with its longer 'burn-up', is usually much more radioactive than metal Magnox fuel. When the Windscale plant was modernized in the early 1960s, with the construction of the chemical separation plant B205, the original reprocessing plant in building B204 was converted into a 'Head End Plant' to prepare oxide fuel for reprocessing. During their sojourn in the reactor the pellets of oxide fuel swell, and wedge themselves inextricably inside their tubular cladding. Accordingly no attempt is made to strip off the metal cladding mechanically. In the Windscale Head End Plant, an entire fuel element was fed gradually into a cave containing an awesome ram-powered shear. This shear would chop through the entire element consisting of perhaps more than 100 fuel pins. Each chop produced a burst of pulverized pellets and a barrage of centimetre-long rings of cladding which dropped into nitric acid. The acid dissolved the remnants of the pellets. The rings of cladding were left behind to be stored, with the Magnox cladding, indefinitely. The acid stream then passed on to the chemical separation plant in the adjacent building B205.

In September 1973 an unexpected chemical reaction in the Head End Plant led to a leakage of radioactivity which slightly contaminated thirty-five employees; the plant was shut down and eventually abandoned. The accident was caused by tiny granules of fission products, insoluble even in nitric acid and intensely radioactive, accumulating in a process vessel. The handling of these granules is one of several technical problems which arise in the reprocessing of high burn-up oxide fuel. The fierce radiation from fission products tends to tear apart the molecules of solvent, especially in the first stage of chemical separation; and the replacement and maintenance of key components, like the shear for chopping up the fuel, which operate in parts of the plant filled with searing radiation, continues to raise questions about the cost and feasibility of reprocessing oxide fuel. The operators of the Windscale works, British Nuclear Fuels Ltd (BNFL) put forward a proposal in the mid 1970s to build a large new 'Thermal Oxide Reprocessing Plant' (THORP). The resulting controversy was prolonged and bitter (see pp. 160–62). Several attempts to reprocess oxide fuel have proved unsuccessful. The only civil oxide fuel reprocessing plant presently in operation is the French UP-2 plant at Cap la Hague; but the Cap la Hague plant too has run into technical difficulties.

Regardless of whether spent fuel is to be reprocessed or not, it must be stored for a time after discharge from the reactor, to allow the short-lived fission products to decay. As we have noted, the usual procedure had long been to discharge the spent fuel into a water-filled cooling pond. Recently, however, partly as a result of growing doubts about the desirability and feasibility of reprocessing, various methods of longer-term storage of spent fuel have been developed. At the Wylfa Magnox station of the CEGB in Wales, spent fuel is discharged not into a pond but into a storage magazine cooled by carbon dioxide – exactly the same environment as the fuel experiences inside the reactors. Magnox cladding deteriorates rapidly in contact with water, becoming unreliable after a year or so; but it can be left in a dry gas-cooled storage magazine more or less indefinitely. The CEGB has recently added two more dry storage magazines at Wylfa, cooled by natural circulation of ordinary air. The stored fuel is reported to be still sound and safe in storage after some years, and experts conclude that such a storage facility would be suitable not only for gas-cooled but also for more hardy water-cooled fuel, for periods extending into decades if desired. They note also that the longer fuel of any kind is stored, the lower is its remaining radioactivity; reprocessing, if eventually undertaken, would therefore be easier the longer the fuel had been stored.

Radioactive Waste

Throughout the nuclear fuel cycle, the materials involved share one common property: they are all to some extent radioactive, that is, they emit radiation. Natural radioactive materials are encountered in mining and milling; the materials remain radioactive throughout enrichment, fuel fabrication and transport – but their activity is not particularly intense. This, however, changes dramatically once they find themselves inside an operating reactor: neutrons from reactor cores tend to make their entire neighbourhood radioactive. So long as the materials in this neighbourhood remain within the biological shielding all is well; but radioactivity inevitably finds a number of escape routes from the confines of reactors, however well buttoned up. The most important of these is via refuelling, when the entire radioactive inventory of the spent fuel is removed from the core of the reactor. We shall discuss in a moment the eventual fate of this concentrated or 'high-level' radioactivity. But less concentrated radioactivity also makes its way directly out of an operating reactor, and must be dealt with.

Low-level Waste

Any radioactivity which emerges into the environment outside the biological shield in the course of routine reactor operation is called a 'running release'. The simplest kind of running release originates just inside the biological shield itself. In reactors with concrete shielding close to the core, it is desirable to keep the concrete from being exposed directly to the heat of the core. Accordingly, except in the case of prestressed concrete pressure vessels, a thin layer of air is blown up the inside wall of the concrete and discharged to the atmosphere from a stack atop the reactor building. Some of the nuclei in the air absorb neutrons and become radioactive, a process called 'neutron activation'. The most notable of the 'activation products' is argon-41, a radioisotope of the inert gas argon. Some reactors are known to discharge hundreds of thousands of curies of argon-41 annually. Fortunately, however, argon-41 has a very short half-life, only some 1.8 hours; so this apparently enormous output decays to a very low activity before drifting from the stack down to ground level. Other atoms in the air also become activated, but only in small amounts and/or for very short half-lives.

Reactor coolant may carry radioactivity out of the biological shield. Impurities in water or graphite moderator are susceptible to neutron activation. Carbon – from graphite moderator or carbon dioxide coolant or both – can become radioactive carbon-14. But since normal carbon is carbon-12 the transmutation requires the absorption of not one but two neutrons and is accordingly infrequent. Heavy water coolant can absorb neutrons, turning the deuterium (hydrogen-2) into hydrogen-3, or tritium, which is radioactive. But the one coolant which does respond readily to neutron activation is the sodium coolant in liquid metal fast breeder reactors. As already indicated it becomes sodium-24, so intensely gamma-active that it must be kept entirely within the biological shield.

The fuel cladding too may contribute to the activity in the cooling circuit, as the cladding suffers gradual corrosion by the hot coolant. Again it is primarily a consequence of impurities in the cladding, which has, of course, been made as little susceptible to neutron absorption as possible, for reasons of neutron economy. The worst offenders in this category are impurities in zircaloy cladding on water-cooled reactor fuel. Corrosion of this cladding is enhanced by the intimate contact with the fast-moving fluid at high pressure, which quickly carries surface corrosion into the coolant flow.

Much more serious are leaks from the fuel cladding, to which some reactors seem prone. The build-up of gaseous fission products inside a fuel rod imposes increasing strain on the cladding; if for any reason the cladding develops a flaw the fission gas seeks it out and escapes into the coolant. A more sizeable leak also releases the volatile fission products, among them the dangerous iodine-131 (see pp. 115–16). 'Burst can detection gear' in a Magnox reactor sniffs out tell-tale radioactivity in the coolant, and locates faulty fuel elements. If the reactor can be refuelled on load, it is possible to remove leaking fuel without a shutdown. If the reactor, like most water-cooled designs, can only be refuelled off load, a shutdown would be necessary. Furthermore fuel which is leaking slowly may be difficult to locate. In any case, early replacement of the fuel disrupts the fuel programme and distorts the planned pattern of neutron density in the core. For all these reasons leaking fuel is frequently left in a reactor until routine refuelling.

These effects make it necessary to decontaminate the cooling circuits of a reactor. Otherwise the unavoidable leakage of radioactivity through valve seals and other permeable points becomes a potential hazard to personnel, and may interfere with maintenance. Decontamination is usually done routinely, by bleeding off a small portion of the coolant and replacing it with fresh uncontaminated coolant. Of course, each time a refuelling machine is coupled to a reactor vessel for on-load refuelling the machine acquires a share of the activity in the coolant; this activity must be discharged – and kept track of – when the machine is depressurized. In the case of carbon dioxide coolant the gas bled off – the 'off gas' – is passed through a variety of filters and delay stages.

Boiling water reactors, which pass the primary coolant directly through turbines, are especially prone to leakage of active coolant. One possible procedure in such a case is to provide storage tanks for coolant bled off. In such tanks the coolant activity can be allowed to decay for some months before it is released. Similar hold-up tanks may also be provided for drainage from floors, for water from sinks used in de-contamination, and for water from the laundry in which contaminated clothing is cleaned. Wastes which require interim storage in hold-up tanks or the like are called medium-level wastes. The cooling ponds for storing irradiated fuel prior to shipment for long-term storage or reprocessing usually pick up activity from the exterior of the fuel elements and also from any leaking elements; so this cooling water, too, has to be dealt with. A common procedure is to cycle the water slowly through the ponds, continually diverting a small fraction, mixing and diluting it with the much greater mass of cooling water discharged from the turbine condensers back into the river or coastal water from which it has been abstracted. Water treatment systems may also use ion exchangers and other standard forms of purification installations to collect and segregate radioactive compounds in sludges.

In the course of everyday business in a reactor plant a certain amount of solid material also becomes contaminated with radioactivity – floor mops, paper towels, broken glassware from sampling labs, etc. The volume of contaminated solids ought to be reduced by, say, incineration; but at present they are usually simply buried in designated burial grounds, or dumped at sea in prepared containers, as are active ion-exchange sludges.

All of the radioactivity which reaches the outside world directly from a reactor installation by these routes may be lumped together as 'low-level' radioactivity. Until the mid 1970s it was not looked upon as much of a problem. Since then, however, two aspects of low-level waste have begun to attract concern. One is the sheer bulk of dry low-level waste; adequate dumping sites, away from ground-water, and able to isolate the radioactivity, are proving more and more difficult to find. More worrying is the presence in some low-level wastes, both solid and liquid, of traces of plutonium and other biologically hazardous transuranic elements. Removal of these traces would be troublesome and prohibitively expensive; but they may nevertheless make the wastes too dangerous to release uncontrolled into the environment. Such wastes – in the UK called PCM, plutonium-contaminated material, and in the USA called TRU for transuranic – seem likely to become more of a problem, especially if reprocessing and the fast breeder reactor become more common.

Decommissioning

Just as reactor fuel eventually becomes no longer usable, so too the reactor itself will in due course, for one reason or another, be shut down permanently. Unlike a fossil-fuelled power station, however, a reactor cannot simply thereafter be dismantled and the ground cleared for future use. As indicated earlier, some parts of the reactor – the solid moderator and other core materials in gas-cooled reactors, the pressure vessel and possibly other parts of the primary cooling circuit, possibly also the concrete biological shielding, and possibly also the spent fuel cooling pond – will be contaminated with radioactivity. This complicates both the process of dismantling and the disposal of the resulting remains. The question has recently, albeit belatedly, received a great deal of study. The general consensus is that such 'decommissioning' of a reactor can be carried out in three stages. The first and least expensive stage would involve removal of the spent fuel and drainage of the cooling circuits, leaving all the fixed structure of the power station in place; the reactor building would be locked and the plant isolated from access by the casual public, probably under physical surveillance of some kind to prevent unauthorized entry. The second stage would involve removal by dismantling and demolition of all the fixed structure of the power station except the reactor itself. The third stage would complete the process by removing the reactor itself – pressure vessel, core internals, pipework, steam generators, the lot – and clearing the remaining concrete away to leave a site available for any future use desired: anything from building a new nuclear power station to growing Brussels sprouts.

Such, at least, is the theory of decommissioning. Unfortunately, it remains to a large extent unsupported by practice. No one anywhere has ever decommissioned a large power reactor after its normal working life. Small research reactors have been decommissioned, and so have a handful of experimental power reactors; but the power reactors which have been decommissioned are, not surprisingly, those which did not long remain in service after start-up. Their core materials had received only brief exposure to neutron radiation, and had little opportunity to accumulate radioactive contamination. Even so, the task of de-commissioning them has been demanding.

The first power reactor to be decommissioned after even moderate service was the Elk River reactor in Minnesota, a 22-MWe BWR which operated only intermittently from 1964 until its permanent shutdown in 1968. The first and second stage decommissioning was more or less straightforward; but removal of the reactor vessel itself required some striking imagination. In order to be able to cut apart the heavy steel vessel, without exposing workers to any more radiation than necessary from activation products in the steel, the demolition team filled the vessel with water and lowered divers into it with underwater cutting torches like those normally used on offshore oil installations. Needless to say the entire exercise was not cheap. Whether it would be feasible at all for a much larger reactor, after much more radiation exposure of the materials, is as yet unclear. Further important information about techniques and costs will be gathered from the decommissioning of the Shippingport reactor – the first nuclear power plant in the USA – which began in 1981. Until someone decommissions a full scale power reactor, the problems will remain uncertain – and the costs frankly 'guesstimates'. Some guesstimates have put the cost as high as the original cost of the reactor; others are much more optimistic. It seems fairly safe to assume that no one will be in a hurry to embark even on second-stage decommissioning, much less third-stage, any sooner than absolutely necessary. There will undoubtedly be a good many mothballed nuclear power stations dotted about the landscape in the early twenty-first century.

High-level Waste

Because running releases are as a rule dilute and not very radioactive it is often said that reactors discharge very little radioactivity to their surroundings. This is, strictly speaking, true; but it is slightly misleading. Almost all the radioactivity which leaves a reactor does so within the used fuel removed from its core. Since the reactor has created virtually the whole of this radioactivity, it is somewhat special pleading to imply that the radioactivity thenceforth bears no relation to the reactor. On the contrary: the radioactivity from used reactor fuel is one of the most challenging problems posed by operation of nuclear reactors.

When a fresh fuel element enters a reactor it is as sleek and glossy as a surgical implant. When it emerges again after radiation it is discoloured, possibly even swollen, caked with what the nuclear engineers bluntly call 'crud'. Inside the cladding the fuel now contains unused uranium-235 and -238, plus a wide assortment of other nuclei created by the fission reactions, neutron absorption and radioactive decay: uranium-237, plutonium-239, -240, -241 and -242, americium-241 and other so-called 'actinides', and literally hundreds of different fission-product nuclei and their decay and neutron-activation products, including krypton-85, strontium-89 and -90, iodine-129 and -131, and caesium-137. Some of these species have short half-lives; while the irradiated fuel sits in the cooling pond or travels from the reactor to long-term storage or a reprocessing plant the short-lived isotopes like uranium-237 and iodine-131 decay to insignificance. After, say, a hundred days of cooling, its radioactivity arises mainly from radioisotopes of only about a dozen elements.

If the spent fuel is left in long-term storage the decay processes continue and the radioactivity, still confined, continues to decrease, albeit less rapidly. If, however, the fuel is reprocessed comparatively soon after its removal from the reactor – as is the case for instance with most Magnox fuel – the radioactive contents then follow various different paths to various different destinations. Assuming that the cladding has been gas-tight until it is stripped or chopped open, the gaseous fission products, particularly krypton-85, thereupon emerge into the atmosphere of a hot cell. Krypton-85 has a half-life of about 10·8 years. It is a chemically inactive inert gas, and is accordingly difficult to recapture. Present practice is simply to discharge it from a stack into the outside air. (Until 1971 US authorities regarded the amount of krypton-85 discharged from their reprocessing plants as classified – that is, secret – because it might reveal how much fissile material they had produced.) Radiobiologists do not consider that the consequent gradual build-up of gamma-emitting krypton in the global atmosphere offers any present hazards. But if nuclear power programmes expand as widely as some anticipate, some form of krypton retention will have to be installed in reprocessing plants before the turn of the century. (Liquid air already contains enough concentrated krypton-85 to represent a minor safety hazard to users.) Some similar considerations also apply to iodine-129. It has a half-life of 16 million years, and is therefore not very radioactive; but since it is concentrated, like all isotopes of iodine, in the human thyroid any significant build-up in the environment must be viewed with caution.

Within a reprocessing plant there is also an inevitable accumulation of low-level liquid and solid radioactive wastes exactly like those which collect in a reactor plant – indeed probably more copious. Solid wastes are buried, or dumped at sea, as before. At the Windscale reprocessing plant low-level liquid wastes are discharged into the Solway Firth through twin pipelines emptying under water more than three kilometres off shore, at the rate of some 500 000 litres per day.

All such routine releases of radioactivity are carried out in accordance with standards laid down by national authorities, usually based on the guidelines of the International Commission on Radiological Protection (see Appendix B, p. 232). In the UK, for instance, discharges of radioactive effluent are monitored by the Ministry of Agriculture, Fisheries and Food, as well as by the dischargers themselves, to ensure compliance with standards imposed as conditions of operation in licences issued by, among others, the Inspectorate of Nuclear Installations, under national legislation. As we describe in Appendix B, such standards continue to be the subject of protracted controversy.

If reprocessing takes place, the fuel, apart from the gaseous or volatile fission products and the cladding – and, in some reprocessing plants, including the cladding – is dissolved in nitric acid for first-stage separation. The witches' brew left behind when the uranium and plutonium pass over into the organic solvent is called 'high-level waste'. Without doubt it is the most daunting waste material produced in any industrial process. At Windscale, the reprocessing of one tonne of fuel produces about five cubic metres of high-level waste – that is, about enough to fill five or six bathtubs. The waste contains nitric acid, fission products which are both thermally hot and intensely radioactive, iron from corrosion of plant vessels, chemical impurities from the original fuel, and a dash of carried-over organic solvent. As may be imagined, it requires delicate treatment, to avoid unpleasant side-effects like reactions between solvent and nitric acid at high temperatures. The volume is reduced by evaporation – under vacuum, to keep down temperatures. The procedure must be carried out under careful control – always remotely of course – to avoid crystallization or precipitation where it might prove embarrassing (such as in process lines) and to keep the fission products at a low concentration so that the rate of heat output does not overwhelm the cooling system.

After evaporation the concentrated waste is led to the storage facility near the main reprocessing plant. At Windscale this is a concrete building containing an array of special storage tanks, eight each of 70 cubic metres capacity and six – so far – of 150 cubic metres capacity. The smaller tanks are fitted with cooling coils; each of the larger tanks has seven independent cooling circuits, external water jackets which include leak detectors, and an internal system of agitators to prevent solids from settling on to the bottom of the tank. The cooling circuits on one of the larger tanks can remove up to 2 megawatts of heat – that is, about 13 watts per litre; this in turn limits the permissible concentration of the fission-product stream flowing into the tank. The temperature in the tank is kept about 50°C. Gradual evaporation of the water from the solution is accompanied by gradual decrease of the radioactive heat output; evaporation can be kept in step with heat output to maintain sufficiently low concentration. It is also necessary to prevent a build-up of hydrogen, produced by the breakdown of water molecules by radiation – so-called 'radiolytic hydrogen'. Tanks can be inter-connected, to prevent overloading of cooling circuits with incoming waste of comparatively high output, and to provide transfer facilities in case of a leak. Tanks in use are permanently sealed into massive steel-lined concrete shielding vaults, never to be seen again. A pro-gramme of construction of new tanks keeps spare capacity available. At the end of 1981 the total volume of liquid waste stored at Windscale was about 1000 cubic metres.

Similar tank storage installations are located at reprocessing facilities in the USA, France, Belgium, the Soviet Union, India, China and elsewhere. The most famous – or notorious – is at the Hanford Reservation in Washington state. Here, in 151 very large tanks, is stored the high-level liquid waste – nearly 250 000 cubic metres – resulting from recovery of the plutonium from the Hanford pro-duction reactors, for the US nuclear weapons programme. It is generally reckoned that tank storage of high-level wastes can only allow for a useful life of twenty to twenty-five years per tank, albeit perhaps somewhat longer for tanks of stainless steel. Many of the Hanford tanks are ordinary carbon steel; more than a dozen leaks have already occurred, including at least one very large leak indeed. Between 20 April and 8 June 1973 tank 106T leaked some 435 000 litres of high-level liquid waste into the earth beneath it, while plant personnel continued to pour more liquid into the tank, oblivious to the falling level recorded on measuring instruments. The leak released approximately 40 000 curies of caesium-137, 14 000 curies of strontium-90 and 4 curies of plutonium. Investigators later declared that the radioactivity would not reach the level of the ground-water below the tank; but a drilling programme to locate the hot waste had to be curtailed, lest new drill holes facilitated the downward migration of radioactivity. The leak was the eleventh recorded at Hanford; it was not the last.

Clearly the hazardous life-span of some constituents of high-level waste far outreaches that of a storage tank. Strontium-90 has a half-life of 28 years, caesium-137 one of 30 years. It takes ten half-lives to reduce the radioactivity of a sample a thousandfold ($\frac{1}{2} \times \frac{1}{2} \times \frac{1}{2} \times \frac{1}{2} \times \frac{1}{2} \times \frac{1}{2} \times \frac{1}{2} \times \frac{1}{2} \times \frac{1}{2} \times \frac{1}{2}$ equals $1/1024$). Accordingly, it takes about 300 years for the radioactivity of 1 curie of strontium-90 or caesium-137 to drop to 1 millicurie. The high-level waste from 1 tonne of irradiated fuel includes about 100 000 curies of each. A 1000-MWe PWR produces at least 25 tonnes of irradiated fuel per year – that is, well over 2 million curies of strontium-90 and another 2 million curies of caesium-137. Some 300 years hence this particular contribution will have dwindled by a factor of a thousand, to only 2000 curies of each: except that 2000 curies of strontium-90 is not very 'only'. Multiply such figures by the number of reactors now in operation, under construction or planned, and the magnitude of the consequent problem becomes numbingly apparent. Furthermore present methods do not – for economic reasons, if not for technical reasons – extract all the actinides from the fission product waste; perhaps 1 per cent of the plutonium, with its half-life of 24 400 years, is left behind to add to the unpleasantness of the residue.

It is evident that such quantities of potentially dangerous radio-activity require scrupulous stewardship. While tank storage is regarded as a satisfactory interim measure, efforts continue to devise a long-term solution to the problem. To mitigate this burden somewhat, it is now accepted that high-level waste ought at least to be in solid rather than in liquid form, to immobilize the waste, reducing the possibility of its spreading via leaks or vaporization. Although reprocessing was long considered an essential stage in the management of spent fuel, opinion is now divided as to its desirability. Some consider that spent fuel itself – a solid structure engineered with high integrity to withstand the severe conditions inside an operating reactor – might eventually prove to be the best available waste form for final disposal. It is also pointed out that even should reprocessing prove in due course to be a desirable stage of waste management, it becomes progressively easier the longer the fuel has been stored, because of the decay of the radioactivity. In any case, as we shall describe in later chapters, spent fuel storage is *de facto* an important interim stage in the waste management process, since oxide fuel reprocessing is neither cheap nor readily available.

Be that as it may, a substantial inventory of high-level waste in liquid form already exists and will require appropriate treatment before final disposal. Several approaches for solidification are under development. In the USA, at Hanford, wastes are simply allowed to boil themselves dry inside storage tanks, to be left as solid cake in the tanks. At the National Reactor Testing Station in Idaho, high-level waste is 'calcined' (baked at high temperature) into granules like coarse white sand, which are stored in huge concrete-shielded stainless steel bins underground. Another approach, favoured particularly by the UK and France, is to evaporate and fuse the high-level waste into dense glass – a process called vitrification. The French AVM (Atelier Vitrification Marcoule) came into operation in the late 1970s, producing pillars of borosilicate glass impregnated with high-level waste; British Nuclear Fuels Ltd is building a pilot plant at Windscale based on the French technology. The glass pillars will be stored for the foreseeable future, pending establishment of suitable final disposal facilities in the two countries – and indeed elsewhere, since some of the vitrified waste will eventually be returned to customer countries, under existing contracts.

The favourite approach to final disposal has long been insertion of solidified high-level waste into a stable geological formation. Finding such an ideal formation has proved, however, difficult. For a time rock salt, in beds or domes, was the preferred geological stratum. A hole would be drilled into the floor of an underground gallery in a salt dome. A waste canister would be lowered into the hole – by remote control as usual – and loose salt would be poured in after it. The salt would become soft under the intense heat from the canister and would snuggle close around it, sealing it permanently in place and conducting heat away at an adequate rate to keep the solid waste from melting. But experience has cast some doubt upon the suitability of salt. 'Salt Vault', a pilot scheme for salt-formation storage, was carried out in the USA in the late 1960s near Lyons, Kansas. But, despite earlier official pronouncements to the contrary, the site was eventually abandoned as unsuitable. A company mining salt on a nearby location pumped several thousand cubic metres of water down a drill-hole to bring up dissolved salt; but the water disappeared, casting doubt on the alleged impermeability of the salt formation. Fortunately no high-level waste had yet been buried in it. The search for more reliable formations and locations continues – not without controversy.

It is worth noting in passing that the nuclear industry refers, as a matter of course, to 'waste management'. It looks like a career with a future – a long future.

PART TWO

The World and Nuclear Fission

4 · Beginnings

Purely as a physical phenomenon nuclear fission offers ample scope for intellectual problem-solving. If it implied nothing further it could be left to those specialists who might find satisfaction in its intellectual challenge; the rest of us could busy ourselves with other more pressing concerns. Unfortunately, nuclear fission – as everyone knows – implies much more than abstruse mathematical argument and donnish hair-splitting. Almost from the time it was first recognized, in 1938, nuclear fission has implied not merely articles in learned journals but major decisions of public policy. The social, economic and political context of nuclear fission has been from the beginning an essential factor in its development; in turn, it has exerted an extraordinary range of social, economic and political influence. To foresee with any clarity the shape of the nuclear future, a historical perspective is imperative. It is necessary to know not only how nuclear fission occurs, but also who makes it occur, under what circumstances and for what purposes. We have already alluded, in a preliminary way, to these aspects of the subject. It is now time to step back and examine them in much more detail. As we shall see, nuclear activities from the outset have been characterized by unpredictability and secrecy. Throughout nuclear history, either too little has been known, or enough has been known but too little said.

In 1896, Henri Becquerel discovered radioactivity. Shortly after, by carrying a vial of radium in his pocket and burning himself, he also discovered the most troublesome attribute of radioactivity: its biological effects, actual and potential. An episode which occurred more than fifty years later characterizes the situation that has prevailed since Becquerel's discoveries. Interested organizations were debating the design of an international symbol to convey the warning 'DANGER: RADIATION'. One group of participants, including labour union representatives, favoured the design of a grinning skull with an aura of wavy lines emanating from it. But spokesmen for government and industry groups flatly refused to sanction such a design, which they considered too frightening. As a result the design finally adopted was a circle with three leaves fanning out from it – intelligible only to those to whom it has been explained beforehand, and utterly devoid of prior associations, either benevolent or malevolent.

This is a succinct instance of the schism which divides viewpoints about radiation. As indicated earlier the very biological essentials of the issue are hotly controversial, increasingly so; but we shall confine discussion of them mainly to Appendix B. What is in some ways yet more controversial is the developing social context of radiation, especially that produced by radioactivity. (A good case can be made for concern about other forms of ionizing radiation, especially diagnostic X-rays; but we shall here comment only that diagnostic X-rays should be used only when clearly indicated by medical evidence, and should be generated only by well-shielded apparatus for as brief an exposure as possible.) Before plunging into the tumult of nuclear controversy it is important to stress – since it may later be easy to overlook – that the main health problem is that created by radiation; that the radiation is inherently invisible and detectable only by special instruments; that different forms of radiation present different hazards (see Appendix B); and that the deleterious result of exposure to radiation may not manifest itself for many years. For these reasons it is far from easy to be confident of understanding the effects of radiation. Accordingly, human undertakings involving radioactivity may be peculiarly difficult to evaluate from a public-health standpoint. They may also, as we shall see, be difficult to evaluate by a variety of other criteria, not least the economic.

After Becquerel's discovery came those by Pierre and Marie Curie, who isolated from the uranium ore pitchblende the powerfully radioactive elements polonium and radium. It now appears that the Curies were, paradoxically, fortunate in their poverty. Their laboratory was a draughty attic, and its otherwise undesirable ventilation probably saved Marie Curie from an early death brought on by inhalation of radon from her experimental materials. (The draughty attic did not save her husband, who died of quite another technology, under the wheels of a cart in a Paris street.)

Scientific fascination with the newly discovered radioactive substances was almost at once paralleled by a search for practical application. Roentgen's X-rays were, within months of their discovery, applied in medicine; but within three years the X-rays – which of course required apparatus to generate them – were meeting competition from the radiation of radium and its relatives. Alas for the early successes of radiotherapy: its pioneers, and Marie Curie herself, were among the first to experience the insidious delayed consequences of radiation. So were their patients, some of whom died not from cancer or other afflictions but from radiation burns inflicted with the aim of healing. Radium became for a time a fashionable material. 'Radium spas' were suddenly successful in several parts of Europe, and doctors prescribed medicines containing radium. There was also a vogue for 'luminous dial' wrist-watches; the digits on the watch faces were painted over with a mixture of zinc sulphide and radium, and glowed in the dark. The women who worked in the watch factory used fine brushes to apply this luminous paint: and to give a brush a suitably pointed tip its user would lick it. In due course almost all the 'luminizers' fell sick, with bleeding gums and anaemia, and eventually most developed bone sarcoma – cancer of the bone – from the accumulation of radium in their bodies. A small New Jersey plant alone produced more than forty victims from staff employed between 1915 and 1926.

The radium which now seemed so ubiquitous was being produced from mines like the old silver mine of Joachimsthal, whose uranium ore gave it a new lease of life. But the miners of Joachimsthal, as we have described, were prone to *Bergkrankheit* – lung cancer, induced by inhalation of radon and radon daughters. Medical detective work had by 1930 identified the genesis of this disease, and made clear that it could be prevented only by ensuring thorough ventilation of under-ground uranium mines, a lesson which was to be callously ignored in the American uranium rush of the 1950s. There was, throughout the 1920s, a growing awareness among medical researchers, biologists, and radiation workers themselves that radiation had some unpleasant attributes. But there was, throughout this time, no particular public concern, and no general sense of controversy about radioactive materials and their uses. It was as if the controversy over radiation had an even longer latency period than radiogenic disease.

From 1939 to 1945, there was neither opportunity nor inclination to question the circumstances which might arise in the manufacture and use of radioactive materials: no opportunity, because most of the frantic effort then occurring was under strict conditions of secrecy, and no inclination, because those involved were preoccupied with the much more immediate and pressing fear that Nazi Germany would achieve nuclear weapons-technology first. As all the world now knows, the

Nazis did not. The USA achieved the technology and the weapons, and used them, bringing an abrupt and devastating end to the Second World War.

In the aftermath of the Hiroshima and Nagasaki bombs, the US government set up the Atomic Bomb Casualty Commission. Its function was dual. It was – so far as the Japanese victims were concerned – a source of medical aid for those who had survived the nuclear explosions but had already suffered or might subsequently suffer from the effects of the radiation. It was also – so far as the USA was concerned – an agency which could carry out a large-scale study and documentation of the effects of radiation on human beings. As may be apparent the two roles were not wholly compatible. Many Japanese grew to resent deeply the role which they felt they were being forced to play, as guinea-pigs for the further enlightenment of the world's first and only user of nuclear weapons.

Nonetheless the US medical researchers studying radiation effects also found opportunities closer to home. Fifteen days after the Hiroshima bomb, on 21 August 1945, Harry Daghlian, a physicist at Los Alamos, accidentally allowed a sample of fissile material to reach criticality when he was handling it. His hands and body were raked by a massive burst of radiation, gamma rays and neutrons. Admitted to hospital within half an hour, Daghlian lost sensation in his fingers, then complained of internal pains and finally became delirious. His hair fell out. His white blood cell count surged as his shattered tissues tried vainly to cope. It took him twenty-four days to die.

Daghlian's death brought home to the entire Los Alamos community the grim ethical conflict in which nuclear physicists – the 'atomic scientists' – now found themselves. It is worth stressing, some forty years later, that the first to recognize the dilemma of nuclear energy – the conflict between its constructive and destructive potentials – were the nuclear scientists themselves.

Even before the dropping of the Hiroshima bomb a group of those who had helped to create it signed a memorandum subsequently known as the Franck report, submitted to the US Secretary of War on 11 June 1945, forecasting with dismaying accuracy the nuclear arms race, in the event of the use of the bomb against a military target. James Franck and his colleagues proposed instead that the bomb be demonstrated in a remote site, desert or island, before representatives of Japan and of the allied United Nations, and that the USA renounce its use thenceforth, provided that the rest of the nations of the world agreed to do likewise. But the Franck proposals, as the world knows, fell on stony ground – unlike the Hiroshima and Nagasaki bombs. Later in 1945 the initiators of this report, with colleagues similarly concerned, founded the *Bulletin of the Atomic Scientists*. Since its inception the *Bulletin*, published in Chicago, has remained one of the most perceptive and committed voices addressing nuclear controversy of every kind.

On 21 May 1946 Louis Slotin, a Canadian physicist working at Los Alamos, was performing an exercise he called 'twisting the dragon's tail'. He had done the operation many times, while determining experimentally the details of fast critical assembly of the hemispheres of uranium-235 which had to slam together to produce the desired nuclear explosion. In this way Slotin had determined experimentally the critical mass for the Hiroshima bomb. Slotin's exercise involved sliding the two hemispheres gradually towards one another along a rod, using two screwdrivers, and watching the neutron detectors display the build-up to criticality. On 21 May he was showing the phenomenon to a group of a half-dozen colleagues, when the screwdriver slipped. The room was filled with blue light. Slotin tore the hemispheres apart with his bare hands, and in so doing probably saved the lives of his colleagues. But he himself was doomed, and knew it. He died nine days later, his terminal sickness being meticulously charted by otherwise helpless medical staff. Slotin's Los Alamos colleagues were forbidden for security reasons to alter their daily routine or reveal anything about the accident, while they witnessed his lingering death.

The assembly that killed Louis Slotin was earmarked for the second bomb in a pair of weapons tests which the US Navy was preparing to carry out at Bikini atoll in the Marshall Islands. The Marshall Islands had been before the Second World War a German protectorate; this responsibility was taken over by the USA at the end of the war, but the 'protection' thereafter given sounds reminiscent of the 'protection' referred to by racketeers. Tests 'Able' and 'Baker' were carried out 30 June and 25 July 1946. 'Able' was an atmospheric nuclear explosion, 'Baker' one deep underwater in Bikini's lagoon. Both drew bitter protests from many scientists, notably the Federation of Atomic Scientists, a newly formed federation of local groups of scientists concerned about the implications of their work; subsequently it became the Federation of American Scientists, and in the 1980s is still deeply involved in and outspoken about nuclear issues. It was said that the US Navy staged the Bikini tests primarily to show that the Army was not the only branch of the military with nuclear capability. About 42 000 onlookers arrived in 250 ships and 150 planes – military, media, politicians, diplomats, plus additional thousands of scientists with a piece of the action. Scientists not thus involved insisted that the tests would serve little genuine experimental purpose, and that they would appear as mere sabre-rattling while the UN grappled with the problem of international control of nuclear technology. To them and many others the whole business looked like a grisly public relations exercise to show off the USA's latest accomplishment.

Whatever their *raison d'être*, the Bikini tests had an aspect which the public did not learn at the time. In order to clear the necessary room for their activities, the US Navy in March 1946 unceremoniously evicted 167 Marshall Islanders from Bikini, transporting them to Rongerik, another island many kilometres distant, with much poorer vegetation, soil and fishing, making expansive promises to the evicted islanders which were thereafter quietly forgotten. The Navy repeatedly assured the islanders that all would soon be able to return to their homes. What they did not add was that the 1946 tests played havoc with the fertile lagoon of Bikini, leaving it full of radioactive mud and making marine life for more than 150 kilometres unsafe to eat. Not until 1968 were the first nine islanders permitted to return to Bikini, to an island altered almost beyond recognition by the nuclear explosions and their after-effects. The mangled ruins of US weapons sheds and towers loomed out of the overgrowth. The new vegetation was coarse and unproductive; even the coconut crabs, huge tree-climbing crustaceans looked upon as a Marshall Island delicacy, had accumulated so much strontium-90 in their shells that the islanders had to be forbidden to eat them. Nevertheless by 1980 the islanders themselves had so much radioactivity in their bodies that they were once again removed from the islands. US authorities were forced to admit that the atoll might not be safe for human habitation for a century or more. The first Bikini tests were called, grandiloquently 'Operation Crossroads'. For the Marshall Islanders they must have looked more like a dead end.

No sooner had the Hiroshima and Nagasaki bombs brought a stunning conclusion to the Second World War than the first manifestation of nuclear paranoia became manifest. The bombs had of course been made through a combined effort of US, British and Canadian scientists and engineers; but the Manhattan Engineer District – code name given to the bomb-development project – had all its major facilities in the USA. By midsummer 1945 the Americans had virtually taken over the project, including its direction and, more importantly, the information it generated. Only days after the Nagasaki bomb a bill was presented to Congress whose ultimate effect, as the Atomic Energy Act of 1946 – the McMahon Act – was to make it illegal for Americans thenceforth to give their erstwhile allies any further access to information about nuclear energy. Top-level discussions, including those between the three heads of government, were contradictory in content and inconclusive in outcome. Eventually the three wartime partners embarked on separate programmes. The evanescent hopes for effective international control of nuclear energy were stillborn. Instead, with the American weapons tests at Bikini, beginning in 1946, and the first Soviet atomic bomb test in 1949, the nuclear arms race began. It has seemed ever since to be a race that no one will win.

The McMahon Act was not at the time regarded even by British and Canadian scientists – much less their American colleagues – as a particularly unfortunate step. The scientists, it is true, realized too late the Act's divisive implications. But it was initially hailed as a victory for rationality, in that it specifically overrode any military claim to control of nuclear developments. Instead, the McMahon Act established two civilian bodies to exercise this responsibility and control: the US Atomic Energy Commission (AEC) and the Congressional Joint Committee on Atomic Energy (JCAE). The Act gave the AEC complete control over the funding and direction of post-war nuclear research and development, military and otherwise. The JCAE was to be the Congressional watchdog over the AEC, the channel through which the elected representatives of the public would monitor and oversee American activities in the nuclear field.

The McMahon Act brought an immediate order into the post-Nagasaki nuclear situation, within the USA at any rate. On 1 January 1947 the AEC came formally into being. It qualified for a healthy slice of the Federal budget, and took over all the facilities established for the Manhattan Project; from that date onwards the American nuclear effort assumed a new dimension.

The AEC's fundamental responsibility was of course development of more powerful and efficient nuclear weapons, and provision of the infrastructure to build them in quantity. Undoubtedly the fiercest controversy in the early years of the AEC centred on whether or not to pursue development of a new form of nuclear weapon, long referred to simply as the 'Super'. There is a limit to the amount of fissile material that can be slammed together efficiently into a prompt critical configuration. Accordingly there is a limit to the energy release possible in a pure fission bomb. Since this energy release is equivalent to that of several hundred thousand tonnes of TNT – several hundred 'kilotons' – it might be thought sufficient for most purposes; but Soviet and American weapons designers thought otherwise.

At least some of them did. Robert Oppenheimer, the brilliant wartime director of the Los Alamos Laboratory, thought the Super was ill-advised, and made no secret of his opinion; in due course this was made the basis of one of the shoddiest episodes in American scientific history, the 'trial' of Oppenheimer in April 1954 which permanently deprived him of access to the nuclear information he had been instrumental in developing. The Oppenheimer case underlined an AEC attitude which was to persist long after AEC interest had diversified into civilian applications of nuclear energy. One of the extraordinary powers granted by the McMahon Act permitted the AEC to call upon the services of branches of the Executive – such as the FBI and the CIA. The AEC spent millions every year on exhaustive vetting of employees, nominally for reasons of 'national security'; and the AEC's control of access to information established a pattern that was subsequently difficult to break, even in contexts to all appearances non-military.

The principle of the Super was straightforward. If a fission bomb is surrounded with material containing nuclei of heavy hydrogen – deuterium, or, better still, tritium (hydrogen-3, with one proton and two neutrons in its nucleus) – the ferocious energy of the fission blast speeds up the light nuclei so that they collide and stick together as helium nuclei. Each 'fusion' of two hydrogen nuclei into a helium nucleus releases a burst of neutrons and additional nuclear energy – once again, mass is converted into energy.

Since there is no immediate upper limit on the amount of 'fusible' material that can be so triggered, the energy release of a fission–fusion or 'thermonuclear' bomb – more commonly known as a hydrogen bomb – is effectively unlimited. Further improvements – if that is the correct word – are also possible. The fusion reaction, like the fission reaction, releases free fast neutrons. Accordingly, a triple-decker bomb can be made: a fission bomb surrounded by fusible hydrogen surrounded by ordinary uranium – much cheaper than heavy hydrogen, and unlimited by criticality considerations. The outer layer of uranium intercepts the barrage of neutrons from within it and undergoes fission, adding yet more energy to the total – and, incidentally, adding also a further enormous quantity of fission products, far more than those from the small fission 'trigger'.

Through the late 1940s and into the early 1950s one overriding concern dominated nuclear controversy: the accelerating arms race between the USA and the Soviet Union, and the pursuit of the fusion weapon. Espionage, counter-espionage and the Cold War climate made nuclear secrecy and nuclear secrets a fountainhead of collective paranoia. Whatever the effect of the McMahon Act within the USA, the erstwhile partners of the USA took it as an act of betrayal. The fury and resentment it engendered still linger in the upper echelons of the UK and Canadian nuclear communities. Even the official US announcement of the Hiroshima bomb was regarded in the UK as taking too much of the credit (if such it was) for the USA; on 6 August 1945 the UK Prime Minister issued a stiff statement pointing out the key roles played by the UK and Canadian contributors to the Manhattan project. However, once the initial grievance had been at least suppressed, the UK and Canadian governments reacted quite differently. The Canadians decided that Canada neither wanted nor could build nuclear weapons. The UK scientists who had been engaged at the Montreal laboratory of the wartime project were recalled to the UK – indeed the Canadians felt somewhat as though, having constructed substantial installations, they were being left holding an expensive and useless bag. We shall return to the Canadian effort shortly.

In the UK, the McMahon Act rankled deeply, and it was never seriously doubted that the UK must thereupon embark on her own nuclear-weapons programme. The UK public – and that includes almost all of Parliament – knew, it must be said, virtually nothing of this. Only a passing reference on 12 May 1948, in a House of Commons answer by the Minister of Defence, gave any indication of the furious activity then under way: 'Research and development continue to receive the highest priority in the defence field, and all types of weapons, including atomic weapons, are being developed.' That was all; the Minister would not elaborate, since it was 'not in the public interest' to do so. The organization given the task of developing UK nuclear weapons was the Division of Atomic Energy Production, Ministry of Supply, eventually to become the United Kingdom Atomic Energy Authority (UKAEA).

In only two and a half years the Production Division had completed the Springfields uranium and fuel fabrication plant; the first Windscale pile loaded with Springfields fuel went critical in July 1950. At this time the construction of the reprocessing plant had not even begun; the first irradiated fuel entered the reprocessing plant in late February 1952. On 3 October 1952 the first UK nuclear bomb vaporized the frigate *Plym* in the waters of the Monte Bello Islands just off the north-western coast of Australia

Several nations besides the UK, the USA and Canada had an early foothold in nuclear matters. Germany, Poland, Hungary and other eastern European countries were the origins of many of the scientists who took their abilities to the UK and the USA after the advent of the Nazis – and, to be sure, of some who did not. French scientists participated in the wartime deliberations that led to the Manhattan project. Norway had the Vemork heavy water production plant – until it was blown up by Norwegian partisans in 1943. The Soviet Union was keenly interested in nuclear matters well before the Second World War.

Of these nations the first to embark on a serious nuclear research and development programme was the Soviet Union. Like every other nation then and since, the Soviet Union regarded nuclear matters as government matters, not to be left to industry or academia. In 1943, even as invading Germany was well inside Soviet borders, the Soviet government set up a nuclear weapons research institute in Moscow, directed by Igor Kurchatov, and later to bear his name. The Soviet nuclear programme was fully as intense as that of the Americans, leading to a fission bomb in August 1949, and a thermonuclear bomb four years later to the month.

In 1950, when President Truman gave the go-ahead for US development of the Super, another huge AEC facility was established: the Savannah River complex in South Carolina, with more plutonium production reactors (this time moderated by heavy water), a full-fledged reprocessing plant, waste storage, the lot. But the Super remained elusive. The Americans are commonly credited with having detonated the first thermonuclear explosion on Enewetak in the Marshall Islands, on 1 November 1952; but it was in no sense an 'H-bomb'. It was an explosion of a large-scale experimental installation, nearly 60 tonnes of delicate equipment; it could no more be dropped on an enemy than could an entire factory. The Soviet thermonuclear explosion of 12 August 1953 was a true H-bomb, portable and droppable. At least a sizeable part of the AEC had other things on their minds when on 8 December 1953 President Eisenhower delivered a major address to the UN proposing a programme of 'Atoms for Peace'.

On 1 March 1954, in the long-suffering Bikini atoll in the Marshall Islands, the US detonated its first H-bomb, designated Castle Bravo. They expected an explosion equivalent to 7 million tonnes – megatons – of TNT. They got one equivalent to 15 megatons. An American destroyer found itself in the path of the radioactive dust; its crew responded by carrying out radiation drill, battening hatches, stationing all hands below decks and waiting until fixed hoses had cleansed the contamination off the exterior surfaces of the ship. But no one had told 236 inhabitants of Rongelaap, Rongerik, and Uterik in the Marshall Islands, and twenty-three Japanese crew members of a fishing vessel called the *Fukuryu Maru*, anything about radiation drill.

Uterik, Rongerik and Rongelaap are about 160 kilometres east of Bikini. But wind, in a direction not anticipated by the weapons testers, carried the bomb debris all the way to the other three islands. On 11 March the AEC issued a press statement:

> During the course of a routine atomic test in the Marshall Islands twenty-eight United States personnel and 236 residents were transported from neighbouring atolls to Kwajalein island according to plan as a precautionary measure. These individuals were unexpectedly exposed to some radioactivity. There were no burns. All were reported well. After the completion of the atomic tests, the natives will be returned to their homes.

Roger Rapoport, a perceptive American reporter, has noted drily (see Bibliography), that the evacuations were indeed according to 'plan', but that the plan was not devised until after the accident occurred:

> The victims sustained beta burns, spotty epilations of the head, skin lesions, pigment changes and scarring. And many of the natives did not feel well at

all. They suffered from anorexia (appetite depression), nausea, vomiting and transient depression of the formed elements in their blood. Over the next sixteen years twenty-one of the natives on Rongelaap island would develop thyroid abnormalities and thyroidectomies would be conducted on eighteen of them. All but two of the nineteen children who were less than ten years old when the accident happened developed thyroid abnormalities; and two of them were dwarfed for life.

For three years after Castle Bravo, Rongelaap remained too radioactive for the islanders to return.

The intervention of the US Navy helicopter service to Kwajalein undoubtedly served a double purpose so far as the stricken Marshall Islanders were concerned. It brought them much needed medical attention; it also ensured that this attention was not accompanied by other attention potentially more embarrassing to the weapons testers. The US Navy did not know, however, that its patrol aircraft had overlooked a Japanese fishing boat, the *Fukuryu Maru*. The *Fukuryu Maru* was trawling for tuna east of Bikini on 1 March, beyond the perimeter of the delineated test zone, when it suddenly seemed to the fishermen as though the sun were rising in the west. Within a few hours the boat was dusted with white ash, sifting down on to the decks, and over the hair and clothing of the crew. By evening two of the crew were vomiting, and overcome by dizziness. By 3 March others were suffering similar symptoms, with aching eyes and itching skin. There was clearly something wrong. The fishing boat turned and made for its home port of Yaizu. It arrived a fortnight later, with all hands suffering from radiation sickness, and the boat still contaminated with radioactivity. Six months later some of the crew were still in hospital. On 23 September 1954 the radio operator, Aiticki Kuboyama, died.

The fate of the *Fukuryu Maru* made headlines all over the world. The irony of its name – translated, it means *Lucky Dragon* – gave an additional twist to the grim saga. The Japanese had been the first victims of the atomic bomb; now it could be said that a Japanese was the first to die from the effect of a hydrogen bomb. The traumatic jolt of the *Fukuryu Maru* incident reinforced the profound psychological revulsion with which the Japanese regarded nuclear energy. Some four decades after the Hiroshima and Nagasaki bombs, and three decades after the *Fukuryu Maru*, the Japanese distrust of nuclear energy remains as deep-seated as ever.

The radioactive return of the *Fukuryu Maru* alerted the world, with stunning impact, to the phenomenon of 'radioactive fall-out'. Only a month later, in April 1954, India called for a standstill on nuclear weapons tests; needless to say the call drew little response. In 1955, the UN General Assembly adopted a resolution establishing a scientific committee to inquire into the effects of radiation and nuclear tests, the UN Scientific Committee on the Effects of Atomic Radiation (UNSCEAR). But the jockeying between the nuclear-weapons testing powers continued unabated, as did the tests and the fall-out, and the efforts of scientists and the public to bring some sense into the proceedings. Bertrand Russell drafted an appeal, co-signed by Albert Einstein two days before his death, which became known as the Russell–Einstein Manifesto. It called upon scientists of all nations to unite to seek a way out of the impasse to which nuclear discoveries had brought mankind. Professor Joseph Rotblat, who had left Los Alamos when it became clear that Nazi Germany could not manufacture nuclear weapons, undertook the organization of such an unofficial scientific conference. An American millionaire agreed to play host to the first gathering, at his birthplace in Pugwash, Nova Scotia. The Pugwash movement grew from this first meeting into one of the most effective and influential international avenues of contact between leading scientists of the USA, the Soviet Union, and other countries; its central objective was – and remains – to devise ways to control nuclear developments, to reduce the unparalleled threats they entail.

By 1956 the AEC was willing to concede that if an animal – say a cow – eats grass sprinkled with strontium-90, chemically similar to calcium, not only its bones but also its milk contains the radioisotope. The AEC belatedly agreed that milk was much the most significant source of strontium-90 in the human diet. In 1957 the AEC Biological and Medical Advisory Committee concluded, contrary to earlier AEC assertions, that fall-out from nuclear tests to the end of 1956, far from being devoid of genetic significance, already seemed likely to produce between 2500 and 13 000 major genetic defects per year in the global population. Meanwhile scientists outside the AEC were pinpointing troublesome radioisotopes which the AEC had either failed to take seriously or ignored completely – radioisotopes like carbon-14 and iodine-131. AEC fall-out surveys, it became evident, usually missed iodine-131 completely, because of its short eight-day half-life, despite the fact that iodine-131 would be concentrated, like all iodine isotopes, in the human thyroid, and might therefore do damage (especially to children) far out of proportion to its concentration in the external surroundings.

In 1957, largely as a result of pressure from the public and the scientific community outside the AEC, the US Public Health Service instituted a fall-out monitoring system, which soon expanded into a widespread network taking frequent samples; before long these fall-out monitors had identified strontium-89 and -90, iodine-131 and other hazardous radioisotopes in quantities copious enough to be far from reassuring. Ralph Lapp's classic study *The Voyage of the Lucky Dragon* demonstrated that the AEC was anything but scrupulous in its stewardship of the public welfare.

From the mid 1950s onwards, as a result of the fall-out issue, the world's public began to pay much closer and more critical attention to the activities of the nuclear authorities. Public concern increasingly encompassed not only nuclear weapons but also the mounting official enthusiasm for civil nuclear power.

President Eisenhower's proposal of 'Atoms for Peace' presaged yet a further transformation in the American nuclear scene. A new Atomic Energy Act in 1954 made it possible for private contractors to build reactors and to possess fissile material under licence from the AEC, and declassified a variety of useful information. However the short-term prospects for nuclear electric power generation did not look very enticing to US industry. The technology was certainly promising, and reasonably hard-headed estimates of probable generating costs would have been persuasive, but for the prevailing low costs of oil and – even more competitive – the surging abundance of indigenous natural gas and coal.

There was little doubt that power reactors of several designs could be built, and that they would indeed generate electricity at a cost which was not unreasonable. But only the most sanguine anticipated that power reactors could compete economically in the USA with fossil-fuel generating units before the mid 1960s. On the other hand, it was clear that if this new technology were to establish itself it could not be expected to be economically viable from birth. The AEC accordingly began to draw up plans for a Cooperative Power Reactor Development Project, and US industry began to show tentative signs of interest.

The US superiority in nuclear experience and resources gave them an embarrassment of choices when it came to civil applications, but that same wealth of resources also made the economic context unpropitious. The UK situation was precisely the reverse. Their straitened circumstances meant that they had to focus their efforts within a very narrow range of technology. But the context of post-war European economics, with limited supplies of other fuels available, at moderately stiff prices, meant that power production by nuclear means looked appetizing. By the late 1940s the British Atomic Energy Research Establishment at Harwell was attacking with gusto the design problems posed by power reactors.

In August 1952 the military Chiefs of Staff issued a call for greatly increased production of weapons-plutonium. A power reactor producing plutonium as a by-product would for the nonce have to be regarded as a plutonium reactor producing power as a by-product. Designs for such a dual-purpose reactor had already been undertaken, labelled 'Pippa' ('pressurized pile producing power and plutonium'). Apart from the transposition of the last two items, 'Pippa' was, in March 1953, given the go-ahead. We know it now as the first Calder Hall reactor. In some respects its design was Hobson's choice – neither sufficient enriched uranium nor sufficient heavy water could be guaranteed for otherwise more desirable designs, and the fast breeder did not come into the category of production reactor for short-term purposes. But a long and thoughtful study of the economic status of a natural uranium power reactor was already extant, prepared by R. V. Moore of Harwell in the early autumn of 1950, comparing the performance of a 90-MWe nuclear station with that of a similar coal-fired station (see p. 177). It identified the essential basis of this economic comparison, which was to become a perennial: coal involves low capital costs and high running costs, whereas nuclear power involves high capital costs and low running costs. Moore's analysis showed that the crossover point, at which nuclear electricity from natural uranium became cheaper per unit than coal-fired, was well within attainable nuclear criteria.

In the 1980s some of Moore's assumptions are poignant to recall – an interest rate of 4 per cent, for instance – but the essence of his argument has remained remarkably relevant. Since the world price of oil began its spectacular climb, the attractions of nuclear power generation have been strongly reinforced. The point at which it becomes cheaper, including both capital charges and running costs, to generate a unit of electricity from uranium rather than from oil has shifted markedly in favour of uranium. Environmental and manpower considerations in the coal industry have produced a similar shift in the balance between uranium and coal. Of course such questions of cost are not the whole nuclear story – nor are the questions of cost themselves so readily answerable, as we shall see in Chapter 7.

5 · Reactors Off and Running

Decision-making in nuclear matters has from the outset been the prerogative of government. So long as nuclear activities were directed entirely towards military ends this was clearly appropriate. However, as civil applications of nuclear energy came more and more to the fore, the relationship between government, the scientific community, industry and commerce became progressively more complex. The precise details differed from nation to nation, as we shall see. But the collaborative arrangement developed in a form without effective parallel in any other field of human endeavour, especially as the nuclear fuel cycle, sector by sector, began to exhibit a civil as well as a military side.

The first nation to emphasize the civil rather than the military aspects of nuclear activities was Canada. As the third partner, with the UK and the USA, in the Manhattan bomb-development project, Canada was chosen as the site for the construction of major experimental facilities, including the first reactor outside the USA. But after the Second World War the UK recalled most of her scientists from Canada to undertake the British nuclear programme. The Canadian government could see no point in a Canadian programme of nuclear weapons; this attitude has persisted ever since. Nonetheless, Canada found itself with the beginnings of a nuclear establishment at Chalk River, some 200 kilometres north of Ottawa, whose construction was well under way when the British left.

Canada's first two reactors, built at Chalk River, were the small ZEEP reactor and the considerably larger NRX, which went critical in 1947 and reached its full power of 40 MWt in May 1948. The NRX was a research reactor, in some respects the first precursor of the CANDU. In common with most heavy water designs the NRX was an efficient producer of plutonium, which Canada sold to the USA and Britain, a procedure which has been followed and expanded with later Canadian reactors. The sale of plutonium to the weapons powers helped to finance the first decade of peaceful nuclear research in Canada. But the work at Chalk River focused mainly on using the

NRX for a wide range of basic research. It soon established itself as the most successful experimental reactor in the world. Then, ironically, the one country which after the Second World War had decided to develop its infant nuclear capacity purely for research was the first to play host to a major reactor accident.

On 12 December 1952 a technician in the basement of the NRX build-ing erroneously opened three or four valves (the exact number has never been established) which lifted three or four of the reactor's twelve shut-off rods out of the core. The supervisor, seeing the red lights come on at the control desk, left his assistant in charge and went to see what was going on. In the basement he at once realized what had happened, reset the valves and telephoned his assistant to press buttons 4 and 3 to restore normal operation. In his haste he inadvertently said '4 and 1'; before he could correct himself his assistant had laid down the phone and complied. Unknown to either the supervisor or his assistant, resetting the valves had extinguished the red lights without in fact fully reinserting the rods; the assistant had no reason to question the safety of pressing button 1. Doing so lifted four more shut-off rods out of the core; the power level of the reactor began to rise. Within twenty seconds the assistant realized that all was not well, and pushed the scram button. This should have reinserted all the shut-off rods, but it did not. Only one of the seven or eight withdrawn rods dropped in, and it did so very slowly, taking some ninety seconds to fall just over three metres. The operators decided that it would be necessary to dump the heavy water out of the calandria, a last-ditch emergency provision to shut off the fission reaction. The heavy water took thirty seconds to drain out of the tank, and instruments showed that the reactor power level had fallen to zero.

But in the basement the supervisor and another staff member could see through a doorway water pouring out of the system. They rushed in with a bucket, thinking it was heavy water – but it was in fact light water coolant, and radioactive. Up above there was a rumble and water jetted out of the reactor. Radioactivity alarms began to sound, both in the reactor building and in the chemical extraction building on another part of the Chalk River site. Sirens warned the site personnel to take refuge indoors; a few minutes later came the top-level order to evacuate the entire plant. Only the control room staff remained behind, donning gas-masks. From beginning to end the whole accident sequence had lasted only seventy seconds.

The unintentional lifting of the control rods had let the chain reaction speed up to such an extent that the heat release had melted some of the uranium fuel. The fission energy itself did not produce an explosion. But the heat boiled some of the coolant, forming steam bubbles which were much less absorptive of neutrons, allowing the fission reaction to speed up still more. Within a few seconds the melting uranium fuel and aluminium cladding began to react with water and steam; the hot uranium metal stripped the oxygen from water molecules, leaving free hydrogen; the hydrogen mixed with inrushing air entering through ruptured piping, and the resulting explosion heaved a four-tonne helium gasholder to the top of its travel and jammed it there.

The surge of temperature and pressure, the chemical reactions and the explosions pretty much demolished the reactor core, and disgorged radio-activity in all directions. About 10 000 curies of long-lived fission products were carried into the basement by the leak of 4 million litres of cooling water. Fortunately the emergency procedures at the plant were effective. It was subsequently reported that no staff received excess radiation exposure during the accident itself, and that during the clean-up – a protracted and messy business, lasting many months – the highest dosage received was only 17 roentgens, with most others being below 4 roentgens. Although well above recommended levels these are in the circumstances comparatively modest exposures; when it is noted that the accident sequence included an almost complete failure of the scram-rod system, the reactor staff may even count themselves fortunate.

Needless to say the lingering radioactive contamination presented the most awkward obstacle to those charged with tidying up the shambles. They can in fact be credited with a coup of sorts: they managed in due course to devise a method whereby they could extract the entire con-taminated calandria from the interior of the reactor shielding and replace it with a new structure. Few of those at Chalk River on the evening of 12 December 1952 would ever have expected to see the NRX back in service after its resounding hiccup. But in only 14 months the NRX was back – in plenty of time to take over again later in the 1950s while its 200-MWt successor, the NRU, underwent two months of extensive de-contamination. The decontamination became necessary on 25 May 1958 when an irradiated fuel element broke and caught fire inside the NRU refuelling machine. At one point a 1-metre length of fiercely radioactive fuel fell out of the refuelling machine and burned. Fortunately it had landed in a maintenance pit; the radiation dose-rate in the pit was estimated *as high as 10 000 roentgens per hour. Some 600 men were involved in the clean-up, and 400 000 square metres around the NRU building were contaminated.*

Like the Canadians, the French participated in the early stages of the Manhattan project, and were then gradually eased out of the picture. Like the Canadians, the French, immediately after the Second World War, set up a government nuclear undertaking directed to pure research: the Commissariat à l'Énergie Atomique (CEA). Like the Canadians, the French found themselves with useful reserves of uranium to be mined. Unlike the Canadians, however, the French began to wonder by the mid 1950s whether it might not be advisable to proceed with the development of nuclear weapons.

Their first reactor, Zoë or – less poetically – EL-1, went critical in 1948, at the research centre at Fontenay-aux-Roses near Paris. By 1952 the first French nuclear power programme had been drawn up. The air-cooled, graphite-moderated G-1 reactor at Marcoule, a 3-MWe dual-purpose plutonium-power reactor went critical in 1956, like its larger British cousin at Calder Hall. As in Britain, the first commercial nuclear stations in France were offspring of the plutonium-production reactors. The 70-MWe Chinon-1 station built on the Loire for Électricité de France went critical in September 1962.

After the start-up of the Obninsk 'first atomic power station' APS-1, and with the gradual relaxation of East–West tension, the Soviet nuclear effort also expanded from military into civil applications. The first full-scale Soviet nuclear power station was constructed at Troitsk in south-western Siberia and came on stream in 1958. Its reactors, eventually six in number, were 100-MWe descendants of the Obninsk plant, a distinctive Soviet design called the RBMK, using a graphite moderator enclosing pressure tubes filled with light-water coolant. Work also began on a Soviet design of pressurized water reactor – 'WWR' in Russian. The first 265-MWe PWR started up at Novo-voronezh in October 1963.

The UK Atomic Energy Authority was created on 1 January 1954, and a government White Paper published in February 1955 laid down the basis for a programme of civil nuclear power in the UK. The White Paper pointed out the anticipated growth in demand for electricity in the UK, the inability of the coal industry to keep up with this demand, and the long-term likelihood that nuclear electricity would prove cheaper than coal; it made no attempt to argue that nuclear electricity would be instantly competitive with coal, but asserted that Britain's lead in technology could not be allowed to dwindle. On the basis that the costs would be comparable to those of electricity from coal, the government gave the go-ahead for a programme of twelve nuclear stations to be constructed throughout the ensuing decade. This programme was later twice revised, in October 1957 and then in June 1960, into a less ambitious one. But even before Calder Hall had started up, the Central Electricity Generating Board had begun to order the first generation of commercial Magnox stations, beginning with the twin-reactor stations at Berkeley in Gloucestershire and Bradwell in Essex.

On 8 October 1957 the physicist in charge of the Windscale Number One plutonium-production reactor threw a switch too soon. He was carrying out a routine operation known as releasing Wigner energy, which involved raising and lowering the power level. According to his instruments he deduced that the core temperature was falling, without completing the desired Wigner release. He did not have a Pile Operating Manual, with its special sections on Wigner release, to help him, nor had he sufficiently detailed instructions. The physicist nonetheless decided to give the power level another short boost, to bring the temperature back up and complete the Wigner release. What he did not know was that thermocouples recording core temperature were not in the hottest part of the core. Core temperatures at some points were considerably higher than the physicist realized. When at 11.05 a.m. he withdrew control rods to raise the power level again, the resulting additional temperature rise eventually ignited at least one fuel rod.

The physicist had no idea that anything was amiss. Not until 5.40 a.m. on 10 October – 42 hours 35 minutes later – was there any external sign that all was not well inside the core of the Windscale Number One reactor. Then instruments began to show that radioactivity was reaching the filters on the top of the cooling-air discharge stack. The filters were known as 'Cockcroft's Folly'; Sir John Cockcroft had insisted that they be installed, after the stack had been built, as a precautionary measure – to the derision of some of his colleagues. As it turned out 'Cockcroft's Folly' probably kept a major accident from becoming a catastrophe. By the time the Windscale staff realized that something was wrong, the fire was an inferno, and spreading fast.

Unfortunately it was far from clear what could be done about it.

Molten uranium and cladding, steeped in fission products, burned fiercely in about 150 fuel channels, fanned by the onrush of air which had by this time no hope of cooling the core. The graphite, too, was aflame. Tom Tuohy, later the Windscale General Manager, recalled standing on the pile cap wearing breathing apparatus, looking down through a viewing port above the cooling pond, and seeing flames shooting out of the discharge face of the core and playing against the concrete shielding of the outer wall — concrete whose specifications required that it be kept below a certain temperature, lest it weaken and collapse. At the height of the fire eleven tonnes of uranium were ablaze.

The Windscale staff knew only too well that water and molten metal in contact might react, oxidizing the metal and leaving hydrogen to mix with incoming air and explode. No one could be sure that such an explosion would not rend open the shielding, disgorging a hell-cloud of scorching radioactivity. Some of the staff insisted that they must first try carbon dioxide, despite Tuohy's remorseless reminder that — at the fire's temperature — the oxygen of carbon dioxide would feed the flames as effectively as air.

An ICI tanker-load of fresh liquid carbon dioxide coolant for the Calder Hall reactors had just arrived on site. But Tuohy's prediction proved all too correct; fed with carbon dioxide the flames only intensified. Water was the only recourse. In the early hours of Friday 11 October the decision was made: the Chief Constable of Cumberland was warned of the possibility of an emergency. Firehoses were hauled up the charge face of the reactor. Their nozzles were cut off, and the hoses were instead coupled to the entry ports on a line of fuel channels about a metre above the heart of the fire. By this time the fire had been raging out of control for more than twenty-four hours. Tuohy ordered everyone else out of the plant but himself, one colleague, and the local fire chief. At 8.55 a.m. they turned on the firehoses.

It worked. Slowly the fire subsided and died. But the Windscale troubles were far from over. The staff had wrestled with the fire for more than a day before word reached the press and the public, including local people, that something was wrong at Windscale. Even while Tuohy and his staff were trying to conquer the flames it was clear that the fire had released a vast cloud of radioisotopes from the melted fuel. The stack filters had trapped a large proportion of the escaping radioactivity, but by no means all. Outside the plant the question was — how much radioactivity had belched out of the stack and descended over Westmorland and

Cumberland? Had it gone further? What kind was it, and how dangerous? Most important of all – what had to be done, and done quickly?

One radioisotope above all was rapidly identified as the most hazardous – iodine-131, with its short half-life, high activity and instinct to home in on the human thyroid. (*In due course it was estimated that some 20 000 curies of iodine-131 had been released to the atmosphere.*) Decisions were made. Cattle grazing in fields where the radioisotopes had descended would produce milk laced with radioiodine; such milk must not be drunk. By arrangement between the Atomic Energy Authority, local police, the Milk Marketing Board, and the Ministry of Agriculture, Fisheries and Food, milk from an area of more than 500 square kilometres – some 2 million litres – was instead poured into rivers and the sea. It was said locally that the worst after-effect of the Windscale fire was the sour stench of every waterway for weeks afterwards. Farmers were compensated by the government; it was also said locally that, to judge by the amount of claims, local cattle must have been yielding more milk than any other cattle in the country.

It has never been explained why the government opted for the dramatic gesture of pouring milk away, instead of merely drying it and storing it for a few weeks until the radioiodine had decayed. It must be assumed that the government anticipated a public outcry at any attempt to let 'radioactive milk' return to the market. It is also reasonable to suppose that the government, with the grandiose gesture of pouring the milk away, thought to distract attention from the other possible consequences which might only become manifest many years later. There does not appear to have been any effort to keep track of people who were near Windscale during the fire. It took place more than two decades ago; in the mid 1980s there remains only the lingering local suspicion that a lot of people seem to die of diseases like cancer these days. There may well be absolutely no medical or statistical bases for such rumours. The evidence one way or the other seems not to have been considered worth collecting.

In the aftermath of the Windscale fire the Windscale Number Two production reactor was shut down while inquiries were carried out. The full report of the inquiry was never published; the published version made it clear that design changes to the Number Two reactor to prevent a recurrence of the fire would be prohibitively expensive. Both reactors were in due course plugged with concrete and entombed. Fortunately the Magnox stations – the Calder Hall and Chapelcross military installations, and the commercial plants of the new civil programme – operated at a temperature *high enough to obviate the need for Wigner release. The Windscale fire was a once-in-a-lifetime event. For those involved, many of whom later became senior figures in the British nuclear establishment, once was enough.*

In 1959 Parliament passed the Nuclear Installations (Licensing and Insurance) Act, creating the Inspectorate of Nuclear Installations, responsible for the safety of commercial nuclear power stations and research reactors. From the moment a nuclear power station was proposed, throughout its design and construction, its operating lifetime and ultimately its decommissioning, the Nuclear Inspectorate was to be involved – in effect as a technically qualified representative of the public. The Act also introduced provisions severely limiting third-party liability in the event of a nuclear mishap – scarcely a vote of confidence in the effectuality of the newly established Inspectorate.

While the Atomic Energy Authority was building Calder Hall and Chapelcross it was also building its remote Dounreay installation in the northernmost tip of Scotland; the Dounreay Fast Reactor went critical on 14 November 1959. The first reactors at the CEGB's Berkeley and Bradwell Magnox stations went critical in August 1961. The 32-MWe Windscale AGR, the first of its kind, went critical in August 1962, as the Authority continued to develop different reactors. In later years the Berkeley and Bradwell stations were to prove mainstays of the CEGB system.

In the USA, under a joint project financed by the AEC, the Duquesne Power & Light Company of Pennsylvania acquired the Shippingport nuclear power station, the first nuclear power station in the USA. The Shippingport station – using a transplanted naval reactor – was not expected to pay its way. As indicated in Chapter 4 (p. 116), the USA had an abundance of oil and gas at prices certain to undercut the nuclear costs, but it was determined to stay abreast of nuclear developments in Europe. The fanfare over Calder Hall put many American nuclear noses out of joint.

In 1955 the AEC established a Cooperative Power Reactor Demonstration Program, offering substantial government finance to utilities prepared to join the AEC in building nuclear-powered generating stations. Despite the financial enticements, the utilities' lack of enthusiasm was in the event fully justified. Of the early experimental AEC-backed stations only the mainstream light-water designs – Shippingport, Dresden 1, Yankee Rowe, Indian Point 1, Big Rock Point,

Humboldt Bay, La Crosse – and the little Peach Bottom HTGR survived as long as a decade. All the others were soon shut down, because they did not work, because they cost too much, or because they proved to be untrustworthy neighbours. Enrico Fermi 1 (sodium-cooled fast breeder), Hallam (sodium-graphite), CVTR (pressurized heavy water), Piqua (organic moderator) and three boiling water reactors (Elk River, Pathfinder and BONUS) lasted from criticality to shutdown at most eight years, and in one case (Hallam) only two. In the USA, more than in any other country, the gestation period for nuclear power was protracted and uncomfortable.

Curiously enough it was the EBR-1, the world's first source of nuclear electricity, which set the stage for the nuclear struggles in the US. On 29 November 1955, during a safety experiment, the operator mistakenly inserted slow-acting control rods instead of the scram-rods. In the resulting temperature surge, some fifty per cent of the reactor's highly enriched uranium fuel rods melted into a shapeless clump at the bottom of the containment. It would have been a bad omen for those planning construction of the first fast breeder power station in the US – except that not even Lewis Strauss, Chairman of the AEC, heard anything about the EBR-1 accident.

In a thermal reactor, the fissile material is diluted by so much that is not fissile – uranium-238, structural material, moderator, coolant – that the operating configuration of the reactor fuel is just about optimum for reactivity. Any distortion of the core makes it less rather than more reactive. The same could not be said with confidence of a core devoid of moderator and made of highly concentrated fissile materials, pure uranium-235 or plutonium. The planners had to concede that, so far as they knew, a fast reactor core which melted down might collapse into a yet more reactive configuration – possibly introducing so much additional reactivity that the full insertion of control rods would not suffice to shut off the chain reaction. The result might be a runaway, a 'dis-assembly' – that is, a small nuclear explosion.

The possibility of a runaway could not be discounted. To some of the planners the only course open would be to build the first such reactor of any size in a remote location, to minimize the con-sequences should anything go amiss. The British did this, siting their first sizeable fast reactor on the north Scottish coast. But Detroit Edison had good reason to oppose this view; for they were proposing that the AEC should help them build a fast reactor power station not far from Detroit. They would of course, they said, enclose the reactor in a containment strong enough to withstand any conceivable explosion and confine any release of radioactivity. In any case such an accident was highly improbable.

The EBR-1 meltdown disproved the last comfortable assumption; but some of the most obviously interested parties did not find this out for some months. It was not until 5 April 1956 that the Wall Street Journal put a direct question about the EBR-1 meltdown to Strauss as AEC Chairman, to which Strauss replied that it was 'news to him'. Not until that evening, at Strauss's personal behest, did the AEC issue a press release admitting that the meltdown had occurred. It cannot have been particularly welcome news to Detroit Edison, who were by this time leading a consortium of some thirty-five utilities and manufacturers, the Power Reactor Development Corporation, whose objective was to build Detroit Edison a prototype fast reactor power station. The Advisory Committee on Reactor Safeguards may have taken note of the EBR-1 episode; on 6 June 1956 they submitted a report declaring that not enough was known to guarantee public safety if such a plant were to be operated near an urban centre. Still, nothing loth, the Power Reactor Development Corporation duly filed with the AEC on 4 August 1956 an application for a construction licence for a 60-MWe fast reactor station, to be sited at Lagoona Beach on Lake Erie, near Monroe, Michigan, about halfway between Detroit and Toledo – within forty kilometres of these cities and of Ann Arbor, and within fifty kilometres of Windsor, Ontario, across the Canadian border. The proposed station was to be named after the creator of the first reactor, Enrico Fermi.

The proposed Fermi plant was not greeted with delight by the inhabitants of the near-by cities. If they had known of the doubts expressed by the Advisory Committee on Reactor Safeguards they would have been even less enthusiastic. But the sceptical report was not made public; on the contrary, it was forthwith suppressed.

The first Geneva Conference on the Peaceful Uses of Atomic Energy in 1955 had already tackled the tricky subject of reactor safety. On 6 July 1956, at the behest of the Congressional Joint Committee, the AEC instructed a team of its experts, most of them based at the Brookhaven National Laboratory, to prepare a detailed analysis of the possibilities and public health implications of reactor accidents.

The Committee was concerned lest uncertainty about liability in the event of accident deter utilities from building nuclear stations; we shall have more to say about this in a moment. The analysis, document number WASH-740, entitled *Theoretical Possibilities and Consequences of Major Accidents in Large Nuclear Power Plants*, was published in March 1957. The AEC apparently expected that the report, which stressed the extreme improbability of such accidents, would help to pacify the clamour around the western end of Lake Erie. If so, the AEC's expectation miscarried drastically.

For some reason lay readers were inclined to overlook WASH-740's estimates that the likelihood of a major accident was between one chance in 100 000 and one in 1000 million per year per reactor. Since the estimates were based on an almost total lack of actual operating experience, their failure to impress their readership may be understandable. This 'maximum credible accident', as it was called, occurred under the following assumptions: the reactor was of up to 200 MWe output, its core nearing time for refuelling and accordingly containing its largest inventory of fission products; it was situated fifty kilometres from a city with a population of one million people; the accident envisioned a breach of the reactor containment sufficient to release half the core inventory of fission products to the outside surroundings, at a time when the wind would blow the radioactive cloud in the direction of the population centre. What did catch the eye was the scale of the consequences anticipated should one of these apparently highly improbable occurrences nonetheless occur: in the worst case, 3400 deaths, 43 000 injuries, and property damage of $7000 million.

WASH-740 emphasized that such an accident was a very remote probability indeed. This putative remoteness, however, juxtaposed with the astronomical numbers should it nonetheless come to pass, set the insurance business an unparalleled poser. Precisely this uncertainty had prompted the Congressional Joint Committee to direct the preparation of WASH-740. But they did not wait to see it; shortly after so directing, Committee members Melvin Price and Clinton Anderson put forward a bill to both Houses of Congress which enshrined their names in nuclear annals as the Price–Anderson Act.

The Price–Anderson Act became law in 1957. Its purpose was uncomplicated, as were its provisions. In essence, it amounted to this: 'Private utilities will not build and operate nuclear stations if they may be bankrupted by claims arising from a major reactor accident.

Therefore, let a utility be instructed to purchase from private insurers as much coverage as they will sell, against nuclear third-party liability. Thereafter the government will kick in an additional $500 million from Federal funds. Beyond this total there shall be no further financial liability.' That is, in the event that the $7000 million property damage foreseen as the worst possible by WASH-740 should some black day occur, a maximum of – as it turned out – $500 million from government plus $60 million from private insurers would be available. Claims for the remaining $6440 million would not be entertained.

Notwithstanding WASH-740, the Price–Anderson Act – coupled with the veiled threat that the AEC would itself enter the electricity business – persuaded Commonwealth Edison, Consolidated Edison, Pacific Gas & Electric and other major utilities to take their first tentative steps along the nuclear road. In partnership with the AEC they commenced construction of the first generation of what were to become the world's best-selling reactors, the pressurized water reactor and boiling water reactor. The Dresden 1 BWR went critical in October 1959, the Yankee Rowe PWR in August 1960 and the Indian Point 1 PWR in August 1962.

By the end of the 1950s nuclear experience was no longer the exclusive property of the nuclear weapons nations and their Second World War partners. To be sure, such international traffic in nuclear matters had a rocky start. There was, once upon a time and briefly, a United Nations Atomic Energy Commission. It was set up by the UN General Assembly in January 1946, by the first UN resolution, regarding nuclear energy. It did not, however, last long, or accomplish much. Then on 8 December 1953, as the Cold War eased, President Eisenhower addressed the UN General Assembly, declaring that it was time to create a new international body under the UN, dedicated to the concept of 'Atoms for Peace': the International Atomic Energy Agency.

The approval of a draft Statute for the Agency did not come until 26 October 1956. Knotty questions included the matter of accounting for fissile material, the arrangements by which Agency inspectors were to carry out accounting, and in general the whole inspissated issue of 'safeguards': the guarantees which would assure that 'atoms for peace' stayed that way. We shall have more to say about this critical consideration in Chapter 8.

Other Agency activities, perhaps less crucial but also less controversial, made much better headway. The first Atoms for Peace conference in Geneva in 1955 was followed by another in 1958, and two later ones, in 1964 and 1971, each more successful than its precursor in exchange of technological information and mutual enthusiasm. The Agency headquarters in Vienna became a major clearing-house for international nuclear activities, and a vigorous proponent of nuclear benefits for all.

In December 1957, under the auspices of the Organization for European Economic Co-operation (OEEC), eighteen European countries formed the European Nuclear Energy Agency (ENEA), to further the joint development and harmonization of civil applications of nuclear energy. In due course it established three joint projects: the Dragon project (see p. 56), the Halden boiling heavy water reactor project in Norway, and the Eurochemic fuel reprocessing facility at Mol in Belgium. Exchanges of scientific and technical information were promoted, international working groups of experts were set up to consider specialist areas, and efforts begun to coordinate nuclear legislation and radiological protection. In 1960 the OEEC became the Organization for Economic Co-operation and Development (OECD) encompassing also the USA, Canada and Japan. When Japan became a full member, the ENEA dropped the 'European', to become the OECD Nuclear Energy Agency.

The first major post-war economic partnership in Europe was the European Coal and Steel Community, whose activities laid the groundwork for the European Economic Community. The countries of the soon-to-be EEC decided to establish a joint enterprise, modelled on the Coal and Steel Community, to be called Euratom. However, Euratom's grandiose programme of research, development and supranational administration of nuclear matters came unstuck when confronted with stubborn national interest.

Instead the international aspect of nuclear activity began on an essentially bilateral basis, between pairs of countries who could see mutual advantage in exchange of information and technology. Much of the initial momentum for such exchanges came, oddly enough, from US interests who considered that other countries with energy shortages might provide a sizeable market. It was in part this aspect which prompted otherwise uneconomical investments in nuclear stations within the USA; foreign buyers, it was felt, would have reservations about purchasing a technology not yet established in the domestic US market.

The USA supported foreign nuclear development by supplying enriched uranium at knockdown prices and by technical assistance, some of which subsequently re-emerged in the form of vigorous indigenous development of pressurized water reactors and boiling water reactors in Sweden, West Germany and Switzerland, so that the leading European firms became stiff competition for the American industry. West Germany was the first non-weapons nation to start up a nuclear power station, the 15-MWe BWR called VAK at Kahl in November 1960. Meanwhile the UK sold two Magnox reactors overseas, to Italy and Japan. The Latina Magnox station went critical in December 1962, and the Tokai Mura Magnox station in May 1965. These were the only UK power reactors ever to find foreign buyers. In Sweden the 12-MWe Ågesta power station, using a Swedish design of pressurized heavy water reactor, went critical in July 1963.

Canada and the UK had experienced their first serious reactor accidents at a chasteningly early stage in their development programmes: not so the USA, who managed nearly two decades with only comparatively minor mishaps involving reactors. The EBR-1 Mark II core melt accident was by no means trivial, but it resulted in only minor exposure to personnel. Other accidents included the destruction of the BORAX experimental reactor in 1954, fuel damage to one of the Hanford production reactors, and fuel melting in the Heat Transfer Reactor Experiment, the Sodium Reactor Experiment and the Westinghouse Test Reactor. Damage was in several instances costly, as was clean-up. But all these assorted episodes, while not precisely reassuring in their variety and frequency, were more or less minor.

The first major reactor accident in the USA, when it finally occurred, was not only major but ugly. On 3 January 1961 at 4 p.m. John Byrnes, Richard McKinley, and Richard Legg, three young servicemen, went on duty at Stationary Low-Power Reactor No. 1 (SL-1) at the NRTS in Idaho. SL-1 was a 3-MWt prototype military nuclear power plant. It had been shut down for work on instrumentation, and the control-rod drives disconnected. Byrnes, McKinley and Legg had been detailed to reassemble these drives. This required that the central rod be lifted just ten centimetres and coupled to the remote driving mechanism, a straightforward procedure which the three had carried out many times. No one knows exactly what happened on 3 January 1961. Later reconstruction of the fatal four seconds indicated that the refit had actually

*been completed. Then, for reasons that will remain forever unknown –
thoughtlessness, horseplay perhaps – the central control rod, number 9, was
pulled out of the core. The official report by AEC investigators suggests
that the control rod was stuck, and that Legg and Byrnes tried to heave
it up manually. When it came loose it rose not merely ten centimetres but
nearly fifty. The result was catastrophic. The core almost instantly went
supercritical, the fuel fried itself, and the resulting steam explosion
blasted a virtually solid slug of water at the roof of the reactor. The
reactor vessel rose three metres, right through the pile cap. Legg and
McKinley were killed instantly; McKinley's body was impaled in the
ceiling structure on an ejected control rod plug. Byrnes was cut down
by a withering flash of radiation. Automatic alarm systems brought
emergency squads, but even before they reached the reactor their radiation
dose-meters were reading off-scale, more than 500 roentgens per hour, a
lethal level of radiation. The level inside the reactor building was even
higher, more than 800 roentgens per hour. Nonetheless two rescuers rushed
into the wreckage and dragged out Byrnes. But Byrnes died in the
ambulance on the way to the Idaho Falls hospital.*

*Recovery of the other two bodies from the reactor room was a protracted
and difficult operation, and had to be carried out with remote
handling gear. In the emergency recovery operation fourteen other men
received radiation doses of more than 5 roentgens, some of them considerably
more. All three bodies remained so radioactive that twenty days elapsed
before it was safe to handle them for burial; they had to be buried in
lead-lined caskets placed in lead-lined vaults. Not for many months did the
level of contamination in the SL-1 building fall low enough to permit
investigation of what had happened.*

From 1954 onwards the AEC was exercising two functions, laid down
by the Atomic Energy Acts, which were fundamentally in conflict.
On the one hand it was charged with operating its own nuclear
facilities and promoting others; on the other hand it was also solely
responsible for laying down and enforcing regulations for the safety of
personnel and public *vis-à-vis* nuclear energy and radiation. In March
1961, to forestall more comprehensive action from outside, the AEC
arbitrarily divided itself into two. One section was to cover operating
and promotional functions, the other licensing and regulatory
functions. Although this system operated for thirteen years it came
increasingly under fire. The nominal separation of promotional and

regulatory activities left many onlookers unconvinced, especially after the surge of private nuclear activities that began in 1963.

In December 1963, New Jersey Central Power & Light Company announced that it was ordering from General Electric a boiling water reactor of more than 500 MWe – more than twice the size of any previous nuclear plant; this plant, at Oyster Creek, would be built without federal assistance. It seemed that nuclear power had at last made the economic breakthrough so long prophesied. The Oyster Creek order began a trickle that was to become a flood.

Southern California Edison ordered the 430-MWe San Onofre pressurized water reactor, which became – in March 1964 – the first of the new generation of plants to receive a construction licence. Connecticut Yankee Atomic Power ordered the Haddam Neck PWR; Commonwealth Edison added a second BWR to its Dresden station; Niagara Mohawk Power Corporation ordered the Nine Mile Point BWR; Rochester Gas & Electric ordered the Robert Emmet Ginna PWR; Connecticut Light & Power ordered the Millstone BWR; Consumers Power of Michigan ordered the Palisades PWR. Commonwealth Edison added a third BWR, Dresden 3, twin to Dresden 2, and followed with a whole new station not far away, Quad Cities 1 and 2. By mid 1965 the upturn in the ordering of nuclear stations in the USA was accelerating remarkably. By the time the San Onofre station had been run up to full power, in January 1968, orders for nuclear stations totalled a capacity of nearly 50 000 MWe and the rush was if anything growing more hectic; another 22 000 MWe were ordered in 1968.

Not all these orders met a welcome in the neighbourhoods expected to play host. From 1961 onwards stirrings of reluctance, foreshadowed by the Enrico Fermi 1 opposition, began to emerge elsewhere. Plans for nuclear stations at Bodega Head in California (virtually on top of the notorious San Andreas earthquake fault), at Ravenswood in the heart of Queens in New York City and on the famous California beach at Malibu were all withdrawn after stubborn opposition. However, on 12 June 1961, Supreme Court Justice Brennan rendered the Court's majority verdict in favour of granting the Power Reactor Development Corporation a licence to construct the Enrico Fermi 1 fast reactor station; the decision was by no means unanimous, with Justices Douglas and Black contributing a blistering dissent. But the AEC and the Power Reactor Development Corporation got their go-ahead.

The Enrico Fermi-1 plant duly took shape south of Detroit, and went critical in August 1963. Thereafter, a succession of problems kept it far below its design rating, when it was not actually shut down. Fuel swelling and distortion, sodium corrosion in the core, problems with fuel handling gear, and endless trouble with the steam generators pushed costs sky-high and kept electrical output to a trickle. At last, however, the obstreperousness in the steam generators seemed to have been overcome; operators prepared to start up the reactor. On 4 October 1966 the control rods were inched out of the core, and the temperature of the sodium coolant began to rise. The reactor was kept at a low power level overnight; the following morning, 5 October, power-raising commenced. A valve malfunction occupied the morning; from lunchtime until 3 p.m. the power level was brought up to 20 MWt, with another interruption to deal with a malfunctioning pump. Just before 3 p.m. the reactor operator noticed a neutron monitor sending erratic signals from the core. He switched over from automatic to manual control. When the erratic signals ceased, power-raising recommenced. Five minutes later, at 34 MWt, the abnormality showed up again. Other instruments appeared to indicate control rods withdrawn more than normal, and unusually high temperatures at two points in the core. But before the control room staff could work out what was happening radiation alarms began to sound.

At 3.20 p.m. six scram rods were inserted to shut down the reactor, and plant staff found out where the radiation was coming from, and why the reactor was behaving oddly. Samples of the sodium coolant and the argon cover gas were discovered to be laden with highly active fission products. Clearly – for reasons utterly obscure – part of the fuel in the core had melted.

The implications of this situation were ominous in the extreme. The Fermi reactor was a fast breeder: its core was a compact cylinder only about 73 centimetres high and 75 centimetres in diameter, the size of a bass drum, which was designed to be capable of producing more than 200 MWt – that is, as much as 200 000 one-bar electric fires. To achieve this awesome output its fine fuel pins, 14 700 of them, made of 28 per cent enriched uranium clad in stainless steel, had to be aligned to meticulous tolerances, no more than a millimetre or so apart. Furthermore this configuration had to be maintained at a temperature of over 400°C, while submerged in an upflowing torrent of liquid sodium passing through the minuscule channels between the pins. Any disturbance of the Fermi core geometry could impede the flow of coolant, leading to unbalanced thermal expansions and more distortion.

The core geometry had another crucial characteristic, common to fast breeders. Unlike a thermal reactor, whose fuel is usually arranged in an optimum geometry to maximize reactivity, a fast reactor has its fuel in a configuration which may be considerably short of the maximum theoretical reactivity it can exhibit. If the Fermi core had been distorted and melted, it might thereafter be susceptible to local surges of reactivity, intense hot spots, which could lead in turn to chemical reactions between fuel, cladding and coolant, and even to violent chemical explosions. Such chemical explosions, rebounding in a collapsing mass of highly enriched fissile fuel, might even cause a full-fledged nuclear explosion.

No one at the Fermi plant had any very persuasive idea of what to do. Any attempt to enter the reactor with the usual remote-handling gear might disturb the precarious equilibrium in the ruptured core. One blunt absolute loomed out of the nerve-racking uncertainty: they had better not do anything hasty. By one of those ironies which in nuclear history seem to abound, the appalling outcome some citizens of Detroit had warned about all the way to the Supreme Court had come within an ace of occurring.

When the full dimensions of the accident became clear an alert went out to all local police and civil defence authorities to prepare for emergency evacuation of Detroit and other centres. So, at least, insist some of those who received the alert. Official records now show no evidence of an alert; the only way to reconcile the conflicting stories is to presume that a directive was subsequently issued to expunge all trace of it. Be that as it may, the staff at the Fermi plant let some weeks elapse before their first gingerly venture to investigate the core. Eventually, after nearly a year, they ascertained, through a viewing probe specially made to deal with the opacity of the sodium, that something was adrift in the bottom of the reactor. When the Fermi engineers found out what it was they were furious.

Some years previously, while the Advisory Committee on Reactor Safeguards were still reluctantly mulling over the Detroit Edison proposal, the Committee had devoted much time to the possibility of a core meltdown in such a fast reactor. In particular they feared that a meltdown might allow the concentrated fissile material to collapse into a fast critical assembly. Such a rapid addition of reactivity and the resulting energy release might blow apart the reactor containment and strew its radio-activity over the landscape. This possibility nagged so persistently that the ACRS – when the reactor was in the final stages of construction – demanded a special safety precaution to prevent the collapsing mass of fuel from settling in a concentrated mass in the central volume of the containment under the core. They directed that a metal pyramid be built on the floor of the containment, so that molten fuel would run off its sides and spread out. The engineers constructing the reactor protested vehemently at being compelled to add this odd bit as an afterthought at the last minute. As it turned out, they were vindicated. One of the six zirconium triangles which formed the cover of the 'core-catcher' pyramid was not securely anchored. At some stage the uprush of liquid sodium lifted this zirconium triangle, about 20 centimetres long, and swept it into the precisely aligned core, partially blocking the sodium flow. The temperature of the in-adequately cooled fuel pins soared; they melted, warping other pins, further occluding the coolant passages, in a progressive distortion of the assembly which – fortunately – stopped short of a complete meltdown.

Had it not stopped short, no one is quite sure what would have been the outcome. One possibility was sardonically labelled the 'China syndrome'. The molten mass of highly reactive fuel, generating its own fierce heat and far beyond any hope of cooling or control, might sear its way through all containments and into the rock below the foundations of the reactor, melting, burning and exploding as it went, bound for China. The concomitant outpouring of radioactivity, assuming the accident had opened a pathway to the surroundings, would make the locality a no-man's-land indefinitely.

It did not, however, happen. Not quite.

In the UK, all four reactors of the Berkeley and Bradwell Magnox stations were up to full power by the latter half of 1962. The Traws-fynydd station barely beat Hinkley Point A to full power in early 1965; Dungeness A and Sizewell A followed in early 1966. The Oldbury station, ordered in early 1961, was half-finished by early 1965, roughly on schedule; the giant Wylfa station, ordered in mid 1963, was making slightly slower headway. North of the border the one nuclear station in Scotland, Hunterston A, had reached full power in late 1964. The nuclear contribution to the British electricity supply by this time far outstripped that anywhere else in the world, including the USA.

But the Magnox design seemed to have come to the end of its role. The capital cost associated with the enormous reactors of Wylfa was unappetizing, and more compact designs, using enriched uranium, now looked more promising: in particular the American pressurized water reactor and boiling water reactor, and the UK advanced gas-cooled reactor, of which at that stage only the 32-MWe Windscale prototype existed. By the end of 1964 the CEGB was readying itself to order a second generation of nuclear stations, to follow the Magnox series.

Three separate industrial groups were prepared to tender; two offered versions of the AGR, the PWR in partnership with Westinghouse, or the BWR in partnership with US General Electric respectively. The third group, led by Atomic Power Constructions, offered an AGR design closest to that of the UKAEA, and – by pushing tolerances to the limit – were able to persuade the CEGB and the government that their tender was likely to be the most favourable. On 25 May 1965 the Minister of Power told the House of Commons that the second nuclear power programme would be based on the AGR design. In the summer of 1965 Atomic Power Constructions won the contract to build Dungeness B, the CEGB's first twin-reactor AGR station, next to its reliable Dungeness A Magnox station.

The consequences were traumatic for the UK nuclear industry. Atomic Power Constructions proved unable to cope with either management problems or acute engineering difficulties, and in 1969 folded. Contracts for the Hinkley Point B and Hunterston B twin-reactor AGR stations went to the Nuclear Power Group, one of the other two UK consortia; British Nuclear Design and Construction received contracts to build the Hartlepool and Heysham stations. The UK's impressive head start into civil nuclear power foundered so badly on the AGRs that it has never recovered. The final outcome is still unresolved; we shall describe subsequent stages in Chapter 6.

In Sweden, the 10-MWe Ågesta pressurized heavy water reactor, designed and built by the Swedish firm of ASEA, went critical in mid 1963 and reached full power in early 1964. It was intended as a forerunner to a series of Swedish-designed power reactors of a heavy water design; but the first full-scale successor, the Marviken reactor at Norrkoping, never attained criticality. Successive modifications failed to overcome unfavourable core characteristics, which eventually compelled the builders to abandon the project. A fossil-fuelled steam plant was built to power the turbogenerator, prompting locals to dub the Marviken plant 'the world's only oil-fired nuclear station'. Fortunately for ASEA, they had already developed an indigenous boiling water reactor; the Oskarshamn-1 station was ordered in 1965, began construction in 1966, and went critical in late 1970.

Switzerland, too, took its first steps towards nuclear power with an indigenous design, with similarly unhappy results. The Lucens reactor, the first Swiss power reactor, was an experimental design, a pressure-tube reactor cooled by carbon dioxide and moderated by a tank of heavy water. By a stroke of what proved to be remarkable foresight it was built in a cavern under a hill. On 21 January 1969 a pressure tube burst, severely damaging the rest of the core, rupturing the calandria, and releasing radioactive coolant and heavy water into the containment. When the tube burst automatic interlocks sealed the reactor cavern, preventing the escape of radioactivity to the plant's surroundings; fortunately no personnel were in the cavern at the time. The reactor was a write-off. Decontamination of the cavern was protracted and trouble-some; by the early 1970s it was decided to use the cavern for the storage of radioactive waste – not least the many tonnes of mangled metal and other rubble from the Lucens reactor itself.

By the end of the 1960s there had been major reactor accidents in Canada, the UK, the USA and Switzerland. The reactor types involved had included gas-cooled, water-cooled and sodium-cooled; graphite-moderated, light-water moderated, heavy-water moderated, and unmoderated; unpressurized, pressure-vessel and pressure-tube; experimental, plutonium-producing and power-producing: com-binations of virtually all the major varieties of reactor design. The industry stressed that there had never been an accident in a com-mercial reactor which had resulted in danger to the public. The Windscale Number One reactor fire certainly posed a threat of danger to the public; but it was not, of course, a commercial reactor. The SL-1 accident killed three men; but they were not, of course, members of the public. The Enrico Fermi accident released no radioactivity off-site – not quite. The logic of the industry argument was impeccable. But by the beginning of 1970 the public was nevertheless beginning to show signs of widespread disquiet. Ironically, albeit perhaps obviously, the reactor designs which met with the most concerted opposition were the light water reactors – pressurized water reactors and boiling water reactors – which had just begun to dominate the world market.

6 · The Charge of the Light Brigade

For the first decade of their existence the light water reactors – pressurized water reactors (PWRs) and boiling water reactors (BWRs) – were only two among the many design variations emerging from industry drawing boards. At the end of the 1960s they were edging into a lead that rapidly became virtually insurmountable. Italy and Japan had begun with Magnox stations imported from the UK; Sweden and Switzerland had begun with indigenous designs that proved unsuccessful. By 1970, however, all four nations had changed course definitively in favour of light water reactors, either imported or home-grown. In 1970 France did likewise. Despite a first generation of gas-graphite reactors second in extent only to that of the UK, France dropped the gas-cooled lineage flat. The 1970 order for Fessenheim 1, a 930-MWe PWR, opened the door for a veritable deluge of water reactors. From that time onward light water reactors spread over the world so fast as almost to swamp the other two main lines of development, the UK gas-cooled designs and the Canadian heavy water designs. But the light water reactors' surging popularity with the industry was closely paralleled by their burgeoning unpopularity elsewhere.

It was certainly true, as the industry insisted, that there had never been a major accident in a light water reactor. But there had been some near misses. On 5 June 1970, for instance, operators lost control of the Dresden 2 BWR near Chicago for two hours. Dials went off scale, a pen recorder jammed, another ran out of paper, water levels and temperatures rose and fell and the operators repeatedly contravened the written emergency procedures as they struggled to recover their grip on the recalcitrant reactor. On 8 December 1971 Dresden 3, the twin of Dresden 2, went through a virtual re-run of the performance.

Such incidents rapidly became part of nuclear folklore in the USA. They resurfaced repeatedly in the courts, especially following a startling legislative innovation which came into force on 1 January 1970: the National Environment Policy Act, or NEPA for short. The key feature of NEPA was the requirement that any major development project should file with the newly constituted Environmental Protection Agency an 'Environmental Impact Statement' (EIS), which would identify all anticipated environmental effects of the proposed project. An environmental impact statement was further required to assess possible alternatives to the proposed project, and present a credible case against the project as well as a case in its favour. The courts had a field day ruling on issues arising out of NEPA; for nuclear critics it proved a godsend, giving them at last the leverage to prise apart the embrace of mutual self-interest shared by the AEC and the US nuclear industry.

An historic legal judgement in July 1971 ruled against the AEC and in favour of objectors to the planned Calvert Cliffs plant in Maryland. The judgement declared that NEPA was not a vague testament of pious generalities, but an unambiguous demand for a reordering of priorities in specific decision-making procedures – including, very particularly, those of the AEC. The Calvert Cliffs judgement was a landmark in US judicial history – and indeed in environmental action. Furthermore, it gave a monumental boost to the morale of nuclear opposition groups in the USA and – because of the publicity it received – encouraged the gradually developing liaison between different local groups.

At about the same time, also in the USA, an obscure technical disagreement about reactor engineering had begun to surface and come to the notice of critical citizens. The disagreement centred on certain auxiliary systems on the two main American reactor types, the PWR and the BWR. The auxiliary systems in question were the emergency core-cooling systems (ECCS). A present-day light water reactor of either kind has a comparatively high power density (see p. 37). Suppose a pipe in the primary cooling circuit springs a leak: a pipe breaks or a valve sticks open. The high pressure inside the circuit (up to 150 atmospheres in a PWR) blows almost all the cooling water out through the break very quickly, as water between the fuel pins in the core turns to steam: a 'blowdown', or 'loss-of-coolant accident'. Steam has a much lower heat capacity than water and is much less effective in removing the heat which the core continues to generate. Even if automatic shutdown systems immediately scram the reactor, and stop the fission reaction, the heat output from the accumulated fission products – the so-called 'decay heat' – may be, in the case of a large reactor, well over 200 MWt; this heat output cannot be shut down. Unless some means is provided to remove the heat, the temperature in the reactor core will shoot up with extreme rapidity, until the cladding and fuel softens and melts, and the core begins to collapse. If this happens, the consequences – as we have already indicated in the case of the Fermi reactor – may be a catastrophic release of radioactivity to the surroundings. Accordingly light water reactors are provided with a variety of emergency core-cooling systems (ECCS) designed to operate automatically – and very swiftly – if a light water reactor's primary cooling circuit is depressurized.

All external connections to a PWR pressure vessel are made above the level of the core. A present-day PWR has three ECCS, one 'passive' and two 'active'. The passive system is an accumulator injection system: two or more large tanks above the reactor, connected into the primary piping, and filled with cool borated water under pressure. If the primary circuit is depressurized, the drop in pressure opens valves and the cool water pours into the reactor. The two active systems are a low-pressure system which supplies replacement water if a large break drops the primary pressure drastically, and a high-pressure system which supplies replacement water if a small break leaves the primary pressure high. Both the high-pressure injection system and the low-pressure injection system involve power-operated pumps and valves, which are activated by monitoring instruments responding to abnormal pressures or levels in the cooling circuits.

A BWR has a drywell containment leading down to a pressure-suppression pool half-full of cool water. Early BWRs have a high-pressure core spray system, later ones a high-pressure coolant injection system, which are activated by low water-level in the reactor vessel, by low pressure in the primary circuit, or by high pressure in the drywell (which indicates escape of steam from the primary circuit). If high-pressure injection and the feedwater pumps cannot keep the reactor vessel sufficiently full of water it is fully depressurized, by discharging steam into the suppression pool; a low-pressure core spray then comes into service to spray water from above the core, and a low-pressure reflooding system fills up the reactor vessel from below.

So far so good – if these various emergency core-cooling systems do indeed perform as designed. However, by 1971 some doubts from within the industry were being echoed by technically competent outsiders, notably a group based in Boston, the Union of Concerned Scientists. Questions about ECCS cropped up in more and more licensing hearings; and the AEC decided to hold special rule-making hearings on ECCS. The hearings convened in January 1972 in Bethesda, Maryland, and lasted, with interruptions, well over a year. They generated a record, AEC Docket RM-50-1, 22 000 pages long, plus exhibits even longer, which even by mid-summer 1972 was being wheeled about the hearing room on a heavy-duty dolly. The transcript includes some startling material. It chronicles the representations made by various divisions of the AEC itself; by the four vendors of light water reactors – Westinghouse, Combustion Engineering, and Babcock & Wilcox (PWRs) and General Electric (BWRs); by the consolidated electrical utilities; and by the Consolidated National Intervenors, a coalition of more than sixty nuclear opposition groups from all over the USA, who backed the technical testimony of the Union of Concerned Scientists.

The Bethesda hearings revealed a deep split within the AEC as to the adequacy of ECCS as currently understood. Senior safety experts within the AEC repeatedly went on record as harbouring grave reservations about the effectiveness of ECCS in the event of a loss-of-coolant accident, and about the sketchy basis of data upon which confidence in ECCS was founded. A yet more disconcerting discovery was that senior AEC officials were in the habit of 'censoring' information generated by AEC safety studies if it might prove embarrassing. A series of articles by Robert Gillette in *Science* in September 1972 lent further substance to the accusation that the AEC regularly suppressed in-house information about reactor safety problems which might hamper reactor-builders, and threatened its own employees with the sack if they stepped out of line with the official anodyne AEC view.

Be that as it may, in the spring of 1973 the AEC promulgated the new rules. After all the expenditure of money, time and effort, it was decided that the original AEC acceptance criteria for ECCS were, with minor revision, completely adequate. But the ECCS story did not stop there and the near misses continued to accumulate. Nevertheless orders for light water reactors poured in. After a curious lapse in 1967–8, when the number of operating power reactors in the USA dropped from 22 to 18 – because of shutdowns – the number began to mount, almost all PWRs and BWRs. By 1972, according to the International Atomic Energy Agency, there were 33 operable power reactors in the USA, with an output capacity rated at nearly 15 000 MWe. By 1973 this had increased to 56 reactors, with a capacity of over 35 000 MWe.

The nuclear expansion in the USA was the most dramatic, but the phenomenon was occurring worldwide. By 1973 seventeen countries had 167 power reactors with a capacity of nearly 61 000 MWe. The vast majority were of one or another light water design, giving international flavour to the debate on light water reactor safety which was growing ever more heated within the USA.

On 4 August 1972 the AEC gave a send-off to a major Reactor Safety Study, to be directed by Professor Norman Rasmussen of the Massachusetts Institute of Technology. The $3 million study, funded by the AEC, was carried out by AEC staff and consultants at the AEC offices in Germantown, Maryland, and by AEC laboratories and contractors including Battelle, Oak Ridge, Brookhaven and Lawrence Livermore. Some of its early results were included in WASH-1250, *The Safety of Nuclear Power Reactors (Light Water Cooled) and Related Facilities*, published by the AEC in July 1973; this was a hefty compendium of technical data and policy considerations which was seized upon by the many organizations by this time embroiled in nuclear confrontations, including Friends of the Earth, the Natural Resources Defense Council, Ralph Nader, the Scientists' Institute for Public Information, the Sierra Club, and a rapidly expanding constituency of local groups.

Objectors were becoming assiduous collectors of nuclear folklore. There was Millstone-1, whose condensers corroded and leaked seawater into the primary cooling; Quad Cities-2, which operated with a forgotten welding-rig sloshing around inside the pressure vessel; Vermont Yankee, on which the control rods were installed upside down, and which by a combination of ingenious malpractices was later started up with the lid off the pressure vessel; Indian Point-2, in which a major steam pipe split over half its circumference, allowing leaking steam to buckle the steel liner of the containment for more than 12 metres; Palisades, in which the core support barrel worked loose and played hob with the reactor internals, causing an indefinite shutdown and provoking a $300 million lawsuit by the operators, Consumers Power, against the builders, Combustion Engineering; and so on.

However, the light water reactors were having it easy compared to the UK's advanced gas-cooled reactors. In 1969, amid mounting technical, managerial and financial shambles at Dungeness B, Atomic Power Constructions went bankrupt. The CEGB was forced to take over, and to persuade one of the two remaining consortia to continue to work on the ill-fated station. At the end of 1982 – a numbing seventeen years after it was ordered – Dungeness B at long last went critical. The later AGRs – Hinkley Point B and Hunterston B, followed by Hartlepool and Heysham – likewise suffered severe delays and recurring technical problems, albeit not so devastating as those at Dungeness B. It came as little surprise when in late 1973 – with all the AGRs falling steadily behind schedule, and none yet started up – the CEGB declared that it proposed to abandon the gas-cooled lineage and opt for PWRs. What did come as a surprise – indeed as a thunderbolt – was the scale of the proposed PWR programme: thirty-two 1300-MWe PWRs to be ordered from 1974 to 1983, a programme whose magnitude overwhelmed most onlookers, especially as the UK nuclear industry was in disorder after the AGR debacle.

The government had just agreed to draw together the whole reactor-building industry into one consortium, the National Nuclear Corporation, and to place the management of this consortium in the hands of the UK's private General Electric Company who would also become majority shareholders. The General Electric Company shared the CEGB's enthusiasm for the switch to light water reactors, but together they drastically overplayed their hand. Before long opponents of the plan included – as well as the UK wing of Friends of the Earth – the Parliamentary Select Committee, who published a terse and hostile report in February 1974; the Institution of Professional Civil Servants; other trades unions; many people in the UK nuclear industry, including eventually the architect and founder of the UK industry, Lord Hinton; Sir Alan Cottrell, a metallurgist of international standing and until April 1974 government Chief Scientist; and many well-informed and influential commentators in the media.

The arguments were manifold. Questions included the safety of light water reactors; public approval of nuclear power; the credibility of the claim that the 1300-MWe design was 'proven' – since none had yet operated anywhere; the effect of imported components on the balance of payments; the accuracy of the forecasts of electricity demand which required such a vast programme; and the effect on British nuclear morale if its twenty years of achievement were abandoned in favour of transatlantic technology. The situation was further complicated by the election of February 1974, in which the Heath government – apparently in favour of light water reactors – was superseded by the Wilson government, whose loyalties were different.

Whatever the reasons, and many have been plausibly advanced, the government decision, foreshadowed from June 1974 onwards, was to reject the CEGB's proposals. Instead of PWRs, the UK would build steam-generating heavy water reactors (SGHWRs) – neither so large, nor so many. Authorization was given for just six 660-MWe reactors, four for the CEGB and two for the South of Scotland Electricity Board (who had favoured this design throughout), with a further review promised in 1978. Despite the government's careful wording, stressing that the decision in no way implied an unfavourable reflection on light water reactors, it was a major setback. Coupled with belated international recognition of Canada's impressive 2000-MWe Pickering CANDU station, the UK reactor decision gave a new lease of life to heavy water designs, and made the world predominance of light water reactors seem somewhat less inevitable.

Proponents of light water reactors lost little time in returning to the fray. On 21 August 1974, the draft version of the US Reactor Safety Study was published as WASH-1400, *An Assessment of Accident Risks in US Commercial Nuclear Power Plants*, indicating that such risks were minimal. As published it included a brief summary – in effect a child's guide to reactor safety, in the form of a catechism – plus the main report, plus ten volumes of back-up material, weighing in all something over 10 kilograms, and forming a stack about 30 centimetres high.

On the basis of the technique known as event-tree/fault-tree analysis the report concluded that the probability of a serious accident in an American light water reactor nuclear plant was very low indeed – one chance in a million per reactor per year of an accident killing as many as seventy people, and similar low figures for a wide range of other possibilities. In the 'tree' technique a possible malfunction is identified; all the alternative possible consequences of the malfunction are identified; then all the possible consequences of each of these alternatives are identified, and so on. When such a 'tree' has been developed, with more and more branches of possible sequences of effects, probabilities are assigned to each effect, and the cumulative probability of each outcome calculated. This technique was heavily criticized after the publication of WASH-1400; similar techniques employed in the space programme and other high-technology projects had not been conspicuously successful. But the Reactor Safety Study laid down in detail the assumptions underlying the safety philosophy of the US industry, making it possible to establish with much more precision those areas whose analysis could be considered satisfactory, and those still requiring further attention.

Amid all the uproar, orders for light water reactors poured in. It seemed for a time that only the other two members of the original Second World War partnership would be able to hold out – Canada with the heavy water CANDU family and Britain with the gas-graphite family. Of course, for power production, Britain had developed the gas-graphite reactors almost by default. Despite the scale of the gas-graphite Magnox and AGR programmes, and the parallel development of the SGHWR, the UK was really looking towards the liquid metal fast breeder reactor (LMFBR).

From the earliest days of the UK nuclear effort long-range sights had been set on the LMFBR as the ideal form of nuclear power reactor. The small Dounreay Fast Reactor was followed by the 250-MWe Proto-type Fast Reactor, also at Dounreay, ordered in 1966. But problems, especially with the intricate reactor roof, delayed work on the Prototype Fast Reactor. As a result it was overtaken by its cross-channel rival, the French Phénix fast breeder reactor at Marcoule. The Phénix went critical in August 1973, and reached full power on the final day of a major international conference on fast reactor power stations taking place in London in March 1974, a public-relations coup which did not go unremarked. The Prototype Fast Reactor itself went critical just before the opening of the conference.

The USA, too, was by this time in hot pursuit of the LMFBR, albeit some distance behind the British, French and Soviets. The 16·5-MWe Experimental Breeder Reactor-2 went critical in the summer of 1963, and was run up very gradually to full power, which it at length attained in mid 1969. In August 1972 the AEC and two industrial consortia announced plans for what the AEC was pleased to call the first prototype fast breeder power station – presumably in the hope that people would have forgotten Enrico Fermi-1. A new station was to be built at Clinch River, near Oak Ridge in Tennessee; and it was viewed as the centrepiece of the AEC's longer-term FBR pro-gramme – to which the AEC was by the mid 1970s devoting fully one quarter of the total Federal funding for all energy research and develop-ment in the US: about $500 million per year. The estimated costs of the CRBR soared rapidly from $400 million to $2000 million. The 340 utilities who had collectively agreed to put up $250 million – an average of $700 000 per utility, scarcely a vote of confidence – did not increase their contribution.

The Soviet BN-350 at Shevchenko, on the Caspian Sea, was the first of the new generation of fast reactor power stations to go critical, in November 1972. It also had some troubles, especially with its steam generators; but the Soviet authorities continued work on the BN-600, at Beloyarsk. The French, satisfied that this intermediate step could be bypassed, prepared to construct a 1200-MWe Super-Phénix, at Creys-Malville; and the UKAEA continued to spend more than £30 million per year – well over half its annual power programme budget – on the fast reactor.

However, all was not well with the fast reactor's essential supply technology – reprocessing. In the UK, BNFL had just lost its Head End Plant at Windscale, after an unexpected chemical reaction released radioactivity into the plant and contaminated thirty-five workers (see p. 88). After prolonged uncertainty BNFL at length conceded that they had no intention of reopening the Head End Plant. Reprocessing was also causing trouble in the USA. Ten years earlier, in 1964, General Electric had come up with a new technique for reprocessing irradiated reactor fuel that did not generate copious and troublesome quantities of high-level liquid wastes. The process was based on the chemistry of the volatile fluorides of uranium and plutonium. General Electric was sure the new process, called Aquafluor, would be a breakthrough in reprocessing, compact and efficient enough to be used on reactor sites themselves. In 1968 General Electric began construction of a plant based on the Aquafluor process, the Midwest Fuel Recovery Plant, near Morris, Illinois, south of Chicago. Six years later, in July 1974, after well over two years of struggle, General Electric filed with the AEC a report which admitted that the Midwest plant did not work, and probably never would. Its cost had almost doubled over the original estimate of $ 36 million, but General Electric's report offered little hope that much if any would be recovered. The plant was made of massive concrete walls, far from easy to reorganize; but trials revealed virtually insuperable maintenance problems in areas which would be irrevocably contaminated by high-level radioactivity once the plant was operating. The failure of the plant drove General Electric out of the reprocessing business, and left the US nuclear industry in a quandary, with no commercial reprocessing facilities available.

One other commercial reprocessing plant was under construction, at Barnwell, in South Carolina; but it was falling steadily behind schedule and its costs were climbing at an alarming rate. One small reprocessing plant, Nuclear Fuel Services, at West Valley, New York, had operated, but only for six years from 1966 to 1972, before being shut down permanently. Once it was shut down, no one wanted to accept financial or technical responsibility for cleaning up the radioactive mess remaining. Even in 1983 no one appears to know what will become of the West Valley plant and its embarrassing inventory of high-level waste.

The Midwest plant was not the only nuclear casualty of 1974. In the summer of 1974 the Kernkraftwerk Niederaibach, a 100-MWe prototype pressure-tube heavy water reactor in West Germany, joined the list of defunct prototypes, only eighteen months after criticality and at a cost of DM 230 million. However, in an awkward year for nuclear power nothing could equal the saga of the *Mutsu*.

The NS Mutsu *was the prototype nuclear cargo vessel of the Japan Nuclear Ship Development Agency. Its voyage of September 1974 probably did more to set back the cause of nuclear marine propulsion than anything in the industry's worst nightmares. Nuclear ship propulsion has of course a long history, dating from the* USS Nautilus *of the mid 1950s, including many other nuclear subs, American and otherwise, aircraft carriers and other naval vessels, plus a handful of non-military ventures. None of the latter can exactly be called a triumph. The* NS Savannah, *the first American nuclear cargo ship, found herself a seagoing pariah, unwelcome in almost every port in the world. The Soviet Union, perhaps with this situation in mind, built the* Lenin, *a nuclear-powered icebreaker which was not concerned with ports of call, and gained the long-range advantage shared by nuclear subs. West Germany built the* Otto Hahn, *whose career, if unremarkable, was at least free of scandal. But the Mutsu escapade lowered the tone of the argument to the level of the broadest farce.*

The Mutsu, powered by a PWR, was launched in 1969. She was named after her home port; but her home port was far from fond of her. Local fishermen were deeply suspicious, afraid that radioactive discharges from the Mutsu would damage the fisheries, or at any event make it more difficult for them to sell their catches. The Mutsu was ready for sea trials in 1972, but public opposition prevented her from sailing. People were worried lest anything go wrong with the start-up of the ship's reactor, with possible deleterious effects on the rich scallop fisheries of Mutsu Bay. Lengthy discussions between the local officials and the government led to establishment of a federal fund of 100 million yen to cover compensation if the fisheries suffered; by August 1974 most local officials agreed to let the Mutsu *sail. But the mayor and some of the fishermen were adamant, and blockaded the nuclear vessel with some 250 small fishing boats. Then, on 25 August, a typhoon forced the fishing boats to run for shelter, and the* Mutsu *slipped out of the bay at midnight under auxiliary power and with a naval escort.*

On the high sea 800 kilometres from the coast, the reactor was brought to criticality on 28 August; but as power was raised slightly for testing a radiation leak was discovered. The leak was apparently minor, but since the reactor commissioning was being done at sea the reactor operators had to improvise. Their improvisations delighted newspaper sub-editors all over the world, and the Mutsu *became overnight a household word, in the most embarrassing possible sense. First the operators tried brewing up borated boiled rice, as an impromptu shielding cement; this was mildly successful but not completely; so they resorted to old socks, and the waiting newsmen rejoiced afresh. Clearly it would have been preferable to return to port and carry out the necessary tests and modifications with shore facilities available. But the feelings in Mutsu Bay were running so high that the* Mutsu *crew members feared for their safety if they were to attempt to bring the crippled ship into harbour. So the* Mutsu *floated helplessly off the Japanese coast, while frantic negotiations took place between the stubborn Mutsu fishermen and the nuclear ship developers.*

The ship drifted for forty-five days before it was at last permitted to return, and even then only under the most stringent conditions. The authorities had to agree to find the Mutsu *a new port within six months, to leave the fuel in the reactor, to remove all shore-based nuclear facilities within thirty months, and to turn over to the mayor the keys of the fuel-handling crane; furthermore the government was to provide $4 million, to establish a compensation fund lest rumours of radioactivity injure the shellfish sales and to pay for new public works in Mutsu. As an exercise in Japanese nuclear public relations the* Mutsu *episode – for far less obvious physical reasons, to be sure – picked up where the* Fukuryu Maru *had left off twenty years before.*

However, for thrills and spills it remained difficult to top the land-locked LWRs. After three country-wide pipe-crack shutdowns of BWRs, in September and December 1974, and January 1975, came an episode which won instant nuclear immortality for a humble candle. The candle was in the hand of an electrician in the cable-spreading room under the control room of the Brown's Ferry station in Alabama, which had just become the world's largest operating nuclear station, with two BWRs of nearly 1100 MWe each up to power. At 12.30 p.m. on 22 March 1975 the electrician and his mate were checking airflow through wall-penetrations for cables, by holding the candle next to the penetration, when the draught blew the candle flame and ignited the foam plastic packing around the cable-tray. The electricians could not put out the fire.

The temperature rise was noticed by the plant operator, who flooded the room with carbon dioxide and extinguished the fire beneath the control room – but the fire had already spread along the cables into the reactor building. When erratic readings began to appear on the controls for Unit One the operator pressed the manual scram button. Soon he found there was also a half-scram on Unit Two, which he had not ordered; and the speed of the main recirculating pump was being reduced. Quickly he scrammed Unit Two. Until the two units were scrammed they had been supplying some 15 per cent of the total demand on the whole Tennessee Valley Authority grid; the effect of their sudden removal can well be imagined. The fire continued to burn for seven hours, affecting hundreds of cables. According to the US Nuclear Regulatory Commission the fire knocked out all five emergency core-cooling systems on Unit 1. Repairs to the station kept it out of service for eighteen months, and entailed over $40 million in the cost of replacement electricity output alone.

By early 1975, after a year of dramatic ups and downs, the nuclear industry world-wide found itself assessing its situation – perhaps, as an industry, for the first time. It was by this time one of the world's largest and fastest-growing industrial sectors. A 1974 list of firms engaged in nuclear business included over 1800 names in the engineering category alone. There were the giants: Westinghouse, General Electric (USA), Shell, Gulf, Exxon, du Pont, Atlantic Richfield, Union Carbide, and many other multinationals – and there were smaller firms, down to those with only a handful of highly specialized employees engaged in some esoteric corner of the technology. There were suppliers of concrete, steel of many different kinds, lead, copper, graphite, boron, zirconium, heavy water, sodium, carbon dioxide, argon, helium, ion-exchange resins, filters, plastics, insulation – the list goes on indefinitely. There were suppliers of complete nuclear power stations, of complete nuclear steam supply systems, of turbo-generator sets and other heavy electrical gear, of pressure vessels, heat exchangers, steam generators, pipes, pumps, valves, instruments, electronics, computers, control rods, cranes, emergency systems, stand-by diesels – and on and on. Besides engineering suppliers there were suppliers of services, beginning with basic physics and chemistry research, economic analysis, financing, architecture and engineering design, transport, testing and inspection, radiological protection and safety, security, insurance, and of course marketing and public relations.

But the Brown's Ferry candle could have served as a beacon warning the nuclear industry of rocks ahead. The euphoria of 1974 burned out almost as fast as the cable trays at Brown's Ferry. Energy planners had been convinced that the drastic fourfold increase in the price of oil would work to the benefit of nuclear power, by making it economically more competitive. They were wrong. Instead the entire industrial world plunged into a deep recession, severely depressing industrial activities, and with them the requirements for fuel and electricity. Furthermore, the oil price increase, stoking inflation and raising interest rates, brought with it similar increases in the prices of other fuels and electricity. People began to look for ways to cut down on their use of all fuels and electricity, by minimizing waste and improving efficiency. In consequence the demand for electricity failed to grow the way the planners had expected. After the peak year of 1974, orders for new generating plant of all kinds – nuclear included – tailed off dramatically, all over the industrial world.

The annual conference of the Atomic Industrial Forum (AIF), the world's largest nuclear trade organization, took place in San Francisco in November 1975. It had been billed as a celebration of twenty years of the nuclear business; but it felt more like a wake. General Atomic had just lost its last two orders for HTGRs, with the cancellation of the Fulton and Summit stations; two weeks before the AIF conference GA announced its indefinite withdrawal from the reactor business. GA's vice-president, Richard McCormack, told the conference: 'Front line (nuclear) vendors to the electrical utility industry have yet to make a dollar for certain, after some twenty years of effort.' He added that even the largest industry participants had now to ask 'whether this is a legitimate business to be in'.

Within the nuclear industry the answer to this question was becoming more and more ambivalent. Outside the industry, however, there was by 1975 a growing number of people whose answer was unequivocally

'No'. In the USA a new organization called 'Critical Mass', founded the previous year under the aegis of the consumer advocate Ralph Nader, held a conference and demonstration in Washington DC, which coincided with the AIF jamboree in San Francisco. The AIF conference was treated to film of the 'Critical Mass' gathering, flown express from Washington, as a sort of anti-cabaret, which did little to raise the spirits of the AIF delegates. Suddenly the nuclear establishment found itself preoccupied to the point of obsession with the need for 'public acceptance'.

Outside the USA the problem was already pressing. It had arisen in an acute form as early as 1969, with public opposition on both sides of the Rhine to the French plan to build a PWR at Fessenheim. But the French authorities were unmoved by the objections, and the Fessenheim PWR went ahead as planned. On the other side of the Rhine, however, matters did not proceed so smoothly. The tiny hamlet of Wyhl, in the southwestern corner of the Federal Republic of Germany, was first mentioned in 1971 as a possible site for a nuclear station. In the months and years that followed, the intensity of local opposition steadily mounted; but the opposition had little effect on the politicians or the planners. Meeting followed meeting; petition followed petition; demonstration followed demonstration; 400 farm tractors made it clear that the opposition was not confined to what the politicians discounted as 'leftist radicals'. Official permission was duly granted; construction began on 17 February 1975. On 18 February several hundred people moved in and occupied the site. On 20 February some 700 police broke up the occupation, none too gently. From that point on Wyhl became a *cause célèbre* – and not only in the Federal Republic. On 23 February the site was reoccupied; this time the occupation became a virtual settlement, with living quarters, supply lines and rotas of occupiers. In the ensuing months the political authorities debated, the courts deliberated and the occupation continued. Seven years later, after a remarkable war of attrition, there was still no nuclear station at Wyhl, nor yet the beginning of one; and the Wyhl example had inspired nuclear opposition in many other sites in the Federal Republic, the rest of Europe and even North America.

In Switzerland the proposal to build a nuclear station at Kaiseraugst, less than 10 kilometres from the city of Basel, was opposed by tactics closely akin to those used at Wyhl, and with similar results: in 1983 there seems little likelihood that the Kaiseraugst station will be built.

In France, however, the authorities wasted no time on legal niceties or political debate. Again and again, confronted by public protest about official nuclear plans, the French authorities simply sent in the shock troops of the CRS. Hostilities reached a frantic pitch in July 1976 at Creys-Malville, site of the newly announced Super-Phénix 1200-MWe fast breeder power station, which had been given the go-ahead in March 1976. Protesters in tens of thousands converged on the site to hold a mass demonstration against the project. Roads in a wide radius around the site were patrolled by police, preventing many protesters from even reaching the site. Those protesters who did suddenly found themselves under attack by a virtual army of riot police, firing tear gas and brandishing clubs, striking out indiscriminately at men, women and children. The site became a pitched battle, in the course of which one demonstrator was killed.

The battle of Creys-Malville marked an ugly escalation in the confrontation between nuclear authorities and objectors. It was followed by even more violent battles at the Brokdorf site near Hamburg, in November 1976, and at the Grohnde site near Hannover, in March 1977. Both sites looked like prison camps, ringed with barbed wire and watch-towers. The police used helicopters, tear gas, water cannon, and baton charges. Most of the protesters were local people; but there were also small cadres of protesters clearly equipped to meet violence with violence. The majority of the protesters were as appalled as the rest of the country at the ferocity of the clashes. The spectacle of helmeted police and masked protesters in frenzied combat was splashed on front pages all over the world. Yet another demonstration, even larger, took place in September 1977 at Kalkar, on the German side of the border with the Netherlands, the site of the SNR-300 prototype fast breeder station being built jointly by the Federal Republic of Germany, the Netherlands and Belgium. This time the German authorities took no chances. They massed police in unheard-of numbers. Vehicles were stopped and searched on both sides of the border; one scheduled train was blocked by an armoured car on the track, while six helicopters landed around it and all the passengers were ordered out. There was no violence from the objectors at the Kalkar demonstration. Many thousands of people massed to make their protest, then dispersed. A new discipline had begun to assert itself – one which would from then on make it more difficult for the authorities to claim that the objectors were all extremists and subversives.

In an ironic sidelight, the German news magazine *Der Spiegel* revealed in February 1978 that the head of the Kalkar project, Dr Klaus Traube, had long been under covert surveillance, under suspicion of association with 'terrorists'. The surveillance had been carried out, unconstitutionally, by the Federal Constitutional Police. Even with their illicit bugging they had found no evidence that would have stood up in court; but Traube was summarily dismissed from his position. Within less than a year he had become one of the most outspoken and trenchant critics of the official nuclear policy of the Federal Republic – one whose inside knowledge and technical expertise could not be impugned.

At Wyhl the objectors had been concerned primarily about the possible effect of power station waste heat and low-level radioactivity on the local vineyards. At Kaiseraugst, Brokdorf, Grohnde and a number of other controversial sites the concern was for the possible consequences of a major accident at the plant. At the site of the Lemoniz station in northern Spain another factor came into play. The Lemoniz station was to be built by Iberduero, one of the largest privately-owned electrical utilities in Spain; but it was to be built in the heart of the Basque country. It thus became the focus of bitter opposition from Basque activists, who regarded it as a symbol of what they considered to be their political subjection to Madrid. The Lemoniz plant became the target not only of the largest mass demonstrations against nuclear power ever seen anywhere – sometimes numbering more than 250 000 people – but also of the militant separatist organization ETA. The plant was repeatedly bombed; a guard was killed in a machine-gun attack. In 1981 the chief nuclear engineer at Lemoniz was kidnapped by ETA and murdered; in 1982 ETA murdered his successor. By this time Iberduero was despairing of ever operating the plant; despite many attempts they had been unable to find a solution which would be politically acceptable to the local Basque public and their legitimate representatives, to say nothing of the ETA terrorists. Many expected that the Lemoniz reactor would never go critical.

In France, too, a nuclear station became a target for separatists, this time Breton, who bombed the only French heavy water power station, EL-4 at Monts d'Arée, but did little damage. Other bombers attacked the Fessenheim station and a computer centre in Paris belonging to the French reactor manufacturer Framatome. In 1981 an obscure terrorist group fired six rockets at the Creys-Malville station. Violence had thus by this time begun to cast a shadow also over civil nuclear energy.

One of the strangest and most unsettling episodes concerned the case of a young woman nuclear worker in the US. On the evening of 13 November 1974 Karen Silkwood, an employee at the Cimarron, Oklahoma, plutonium plant of the Kerr–McGee Corporation, got into her car to drive to a meeting with an official of her union and a reporter from the New York Times, well known for his hard-nosed articles about the nuclear industry. Karen Silkwood did not, however, keep her appointment. She was found dead in her wrecked car, which had driven off a straight road at speed. She had promised the reporter and the union official to bring with her a dossier on malpractice at the plutonium plant; but no dossier was found in the wreckage. An accident investigator engaged by the union, the Oil, Chemical and Atomic Workers' Union, declared that marks on the rear of her car suggested that it had been rammed from behind: forced into the ditch? It emerged that Karen Silkwood herself and her apartment were both contaminated with traces of plutonium. How the plutonium got there has never been established. Many other questions remain unanswered about the Silkwood case. Her family sued Kerr–McGee and were awarded several million dollars damages: appeals are still in the courts. The Crescent plant was shut down permanently within a few months after Karen Silkwood's death. Many theories have been advanced as to the truth about the Silkwood case; some of them even suggest that she was killed to prevent the disclosure of malfeasance at very high levels, involving illicit traffic in plutonium. Be that as it may, it now seems unlikely that the full truth behind this bizarre affair will ever be known.

The Silkwood case undoubtedly aggravated the unease of many in the USA who were already none too happy with the nuclear state of the nation. In February 1976 three senior nuclear engineers resigned from the staff of General Electric, because they were unhappy about the safety of nuclear plants. Their resignations made headlines all over the world, as did the resignation of another nuclear engineer from the staff of the Nuclear Regulatory Commission itself, for similar reasons. The four 'dissident' engineers were caught up immediately in fiercely contested campaigns in support of several so-called 'initiatives' – in effect, state referendums on nuclear issues. Probably the most hard-fought of these was Proposition 15 in California. Proposition 15 called for a ban on any further construction of nuclear plants in California pending acceptance by the State legislature that the technology was safe and that there existed a satisfactory way to dispose safely of the radioactive waste. Supporters of Proposition 15 gathered hundreds of thousands of signatures to get it onto the official ballot; opponents of Proposition 15, who included all the electrical utilities and many other industries and businesses, spent several million dollars on television, billboard and other advertising, warning of the dire consequences if the state turned its back on nuclear power. In the event, not only Proposition 15 but all the other 'local initiatives', in states from one side of the USA to the other, were defeated. The nuclear industry viewed the outcome as a democratic endorsement of its activities. Nuclear opponents, pointing to the disparity of financial resources available to the proponents, began to wonder aloud whether even in the heavily legalistic climate of the USA it might be better to forsake the courts and other formal procedures and, like their European counterparts, take to the streets.

The first such manifestation in the USA took, not strictly to the streets, but to the seashore. The Public Service Company of New Hampshire (PSC) was proposing to build a nuclear plant at Seabrook, New Hampshire. The site they had chosen was on a stretch of coast well known for its offshore population of clams. Local people who had their own reasons for objecting to the plant declared solidarity with the clams, in what they called the Clamshell Alliance, to oppose the construction of the Seabrook plant. As usual the case wound its way through hearings of various kinds, but PSC duly received permission to build the Seabrook plant. The Clamshell Alliance thereupon instituted a new and novel form of opposition tactic. They decided on direct action at the Seabrook site; but they did not want a US replay of the ugly scenes at Creys-Malville. Accordingly they organized small groups of people and trained them in the methods of non-violent opposition. The Alliance made its strategy one of strict discipline and self-policing; only those deemed psychologically able to cope with the consequences of confrontation with the forces of law were accepted by the organizers. The Seabrook demonstration of October 1976 was a landmark, of a sort, in the history of the nuclear industry in the USA. As 1414 protesters were quietly arrested and dragged off to jail their example was striking sparks all over the USA – and indeed outside the USA. Even in France and the Federal Republic of Germany, scenes of some of the ugliest confrontations between nuclear protesters and the law, many nuclear opponents had long recognized that violent clashes with the authorities did the opposition cause little good. The example of the Clams – silent but tenacious – soon found many emulators.

While nuclear opponents were forging international link-ups in many parts of the world, governments and nuclear promoters were falling out. In mid 1975 Westinghouse, the world's most successful reactor vendor, announced that it was in difficulty. It had sold some two dozen power reactors to customers in the USA and Europe, under contracts guaranteeing a lifetime supply of fuel at a fixed price; but a fourfold increase in the price of uranium took Westinghouse unawares. It had failed to buy in adequate supplies of uranium to meet its contractual obligations; the sudden price rise caught it flat-footed. In June 1975 Westinghouse told its customers that it did not intend to fulfil its supply commitments. The fuel supply provisions had been, however, a major attraction of the original deal offered by Westinghouse; furious customers responded with twenty-nine lawsuits.

At about the same time it was revealed that the governments of Australia, Canada, France and South Africa, together with the international mining company Rio Tinto-Zinc, had secretly agreed to cooperate to raise the world price of uranium. Westinghouse forthwith issued writs in all directions, claiming that its fuel supply difficulties had been caused by this uranium 'cartel'. A US judge hearing the lawsuits against Westinghouse pointed out that he could in law find in favour of the plaintiffs, but that he would thus in effect be declaring Westinghouse bankrupt. He counselled the parties to seek a less drastic way out. Westinghouse did not get very far with its own lawsuits; but they provided a delay while Westinghouse reached settlements with its disgruntled customers. The affair, which had governments, mining companies, electrical utilities and Westinghouse all at each other's throats for several years, did little to foster a climate of mutual trust and confidence within the nuclear community. It was indeed a foretaste of further high-level unpleasantness soon to engulf the world nuclear scene.

In the mid 1970s the nuclear industry came up with more promising ideas. The German firm BASF for a time proposed to construct a nuclear station right on the site of its huge chemical plant at Ludwigshafen, to provide not only electricity but also process-heat. It would have been the first nuclear process-heat plant; but growing doubts about the safety of a PWR in the middle of a vast chemical complex eventually scuttled the proposal. Another innovation, planned by Westinghouse and Tenneco, to construct floating nuclear power stations, to be moored offshore and linked to users by cable, attracted some attention, but at length failed to leave the slipway – mainly because of the general decline in the market for new electricity plant of whatever kind.

This general decline spurred vendors into a new keenness for export business. A report prepared for ERDA by Barber Associates, about nuclear exports to 'less developed countries', published in mid 1975 as ERDA-52, assessed the 'commercial, economic and security implications', and was anything but sanguine about them. But the export rush was already on. To the chagrin of the US vendors the German firm of Kraftwerk Union (KWU) announced in mid 1975 that it had won the largest nuclear export contract ever: a deal whereby KWU would supply Brazil not only with eight nuclear power stations but also with an enrichment plant and a reprocessing plant. Westinghouse was especially aggrieved, since it had already supplied the first Brazilian plant, at Angra dos Reis. But the aspect of the deal which most bothered less partial onlookers was the inclusion of the two fuel cycle facilities – each of them, willy-nilly, capable of producing potential nuclear weapons material. France too embarked on this new approach to exports, agreeing to supply reprocessing plants to South Korea and Pakistan. The implications of international trade in what came to be called 'sensitive technologies' we shall consider further in Chapter 8. As we shall see, the export deals of the mid 1970s set the stage for a major inter-governmental confrontation. They also unsettled those of the general public who had begun to think about the link between civil and military nuclear activities.

This link was no mere science-fiction nightmare dreamt up by hysterical nuclear protesters. Two official reports, published within a month of each other in September and October 1976, made that abundantly, if unhappily, clear. The first was the sixth report from the British Royal Commission on Environmental Pollution, chaired by Sir Brian (later Lord) Flowers. Entitled *Nuclear Power and the Environment*, it at once achieved the status of a classic, as the 'Flowers Report'. Flowers was himself a distinguished nuclear physicist, and a part-time member of the board of the UKAEA; his colleagues on the commission were likewise variously distinguished. In their report they considered virtually every aspect of nuclear controversy, construing the word 'environmental' very widely indeed. Many nuclear questions they discounted as invalid or at most of minor importance. But they were deeply concerned at the prospect of plutonium becoming an article of commerce, directly or indirectly: 'The dangers of the creation of plutonium in large quantities in conditions of world unrest are genuine and serious. We should not rely for energy supply on a process which produces such a hazardous substance as plutonium, unless there is no reasonable alternative.'

Half a world away, in Australia, the Ranger Uranium Environmental Inquiry, under Mr Justice Russell Fox, published their first report on plans to mine uranium in Australia's Northern Territory. Like the Flowers Commission the Fox Commission was prepared to give a clean bill of health to many controversial aspects of nuclear activities; but its third finding cannot have given nuclear people much solace: 'The nuclear power industry is unintentionally contributing to an increased risk of nuclear war. This is the most serious hazard associated with this industry.' We shall return to this crucial problem in Chapter 8. Both the Flowers Report and the Fox Report, dispassionate and authoritative, gave a new credibility to at least some of the arguments which had long been advanced by nuclear opponents. Even within the world's nuclear establishments dissent and tension were beginning to mount.

On 28 October 1976, US President Gerald Ford, in a major policy statement, declared that henceforth the US government would not consider reprocessing or the use of separated plutonium to be a necessary or desirable concomitant of civil nuclear activities, because of the accompanying risk of misuse of plutonium. The theme was reiterated and reinforced on 7 April 1977 by the incoming President Jimmy Carter, whose position was further supported by yet another high-level report, this one published by the Ford Foundation and the Mitre Corporation in the USA. We shall have more to say about these developments in Chapter 8; but they provided an unexpected gloss on a controversy which became the most far-reaching civil nuclear confrontation the UK had yet seen.

British Nuclear Fuels Ltd (BNFL) proposed in 1974 to build one or more large new reprocessing plants at their Windscale site, to service overseas customers. A garbled story about this proposal appeared in the *Daily Mirror* in October 1975, under the stentorian headline 'Plan to Make Britain World's Nuclear Dustbin'. The story triggered a national uproar. In March 1976 the government duly gave the official go-ahead for BNFL to complete a contract with Japanese electrical utilities, to build a new 'Thermal Oxide Reprocessing Plant' (THORP) partly financed by the Japanese, who would ship 3000 tonnes of spent fuel to the UK. But the deal attracted both local and national opposition, and a demand that the plan be subjected to a public inquiry before proceeding. The government continued to stonewall; then, in December 1976, it was revealed belatedly that BNFL had discovered, two months earlier, a leak under one of its radioactive waste storage bunkers at Windscale. The existence of the leak had been kept secret; when the Secretary of State for Energy, Tony Benn, eventually learned about it he was furious. On 22 December 1976 the government announced that it had after all decided to hold an inquiry into BNFL's THORP proposal; and objectors thronged forward. The inquiry opened in the small town of Whitehaven, five miles north of the Windscale site in Cumbria on 14 June 1977, under the eye of the Inspector, a High Court Judge named Roger Parker. It sat for exactly one hundred days, considered some 1500 documents, and accumulated a transcript taking up more than a metre of shelf space.

BNFL argued that the project would be financially advantageous to the UK; was essential for the UK's own nuclear programme; would recover valuable uranium and plutonium for future use; and was necessary as a stage in the management of high-level waste. They also insisted that the high-level waste from overseas customers would in due course be returned to these customers in solid form for final disposal. Objectors argued that reprocessing of oxide fuel was an unproven technology and that everyone who had tried it – including BNFL with the Head End Plant – had failed, either technically or financially or both. Objectors also argued that the uranium and plutonium would cost more to recover than they were worth; that reprocessing of oxide fuel was not only unnecessary but would positively complicate radioactive waste management; and that long-term dry storage was a preferable approach, given the present uncertain state of knowledge about final disposal. Above all – an argument put forward with particular force by the environmental organization Friends of the Earth – if the UK were to claim that it must recover plutonium for use as fuel, it would set a precedent that would gravely jeopardize international efforts to control the proliferation of nuclear weapons. Other countries – India, Pakistan, Argentina, Brazil, South Korea – could then say that if the UK felt it necessary to separate plutonium for civil purposes, so did they.

However, the argument most strenuously advanced by other objectors was that the discharge of low-level radioactivity from the plant would be a health hazard. The Inspector was clearly unimpressed by this argument – and indeed by all the other arguments put forward by the objectors. The official report of the inquiry was published in March 1978. It essentially endorsed BNFL's original plan throughout; and after two debates in the House of Commons, the THORP plan was given parliamentary blessing. Objectors greeted the Parker Report with fury and disbelief, unable to understand why their arguments had been dismissed, often without mention. The Parker Report polarized the nuclear issue in the UK as never before.

Not only the objectors in the UK were worried about the proliferation impact of the Windscale plan. The US government made a series of representations to the UK government – but received only a cool rebuff. The issue of reprocessing and the civil use of plutonium was creating a polarization not only between 'pro' and 'anti' nuclear factions, but also between national governments, as we shall discuss further in Chapter 8. The deep disagreement was already tainting the atmosphere during the IAEA conference on 'Nuclear Power and its Fuel Cycle', in Salzburg, Austria, in May 1977. Many official US participants had to make last-minute changes to their presentations, to keep them in line with US government policy; they in turn were taken to task by many non-US participants. Only a block away, however, nuclear critics and opponents from many countries had forgathered for a 'Conference on a Non-Nuclear Future', the first major international meeting of what had by now become a full-blooded international movement. The unanimity of outlook and purpose at the non-nuclear conference made a striking contrast with the dissension at the official conference in the next block.

One consequence of the non-nuclear conference was the establishment of an international network, to be called the World Information Service on Energy (WISE). Funding for WISE came from an improbable source. The Danish nuclear opposition organization OOA had designed and copyrighted a simple, vivid symbol: a smiling cartoon sun, in bright red on a yellow background, with above it the words 'Nuclear Power?' and below it the answer 'No Thanks'. The smiling sun symbol soon spread to virtually every corner of the industrial world, and even farther afield. By 1983 it was available in more than forty different languages. Proceeds from the sale of smiling sun badges, stickers, T-shirts, and other manifestations were duly collected and returned to the Danish licensees; and the profits, ploughed back into a foundation, provided financing for the WISE headquarters in Amsterdam, the bi-monthly WISE bulletin, and eventually for a growing number of national network links.

For the world's nuclear industry there was by this time little to smile about. By 1975, as nuclear enthusiasm peaked, countries which had announced forthcoming nuclear programmes had included Australia, Denmark, Indonesia, Israel, Kuwait, Norway, New Zealand, Malaysia, Saudi Arabia and Thailand. However, within three years all of these countries declared that nuclear power development plans had been either deferred indefinitely or scrapped outright. Not only the traditional domestic markets, but also the potential export markets for nuclear technology were dwindling rapidly; and the reactor vendors were fighting tooth and nail for the few possible orders remaining.

By early 1975 South Africa, long a major if controversial customer for nuclear technology, and with a highly developed – and highly classified – nuclear research programme, had decided to build its first nuclear power station at Koeberg, near Cape Town. The leading suitor for the South African hand was a consortium led by US General Electric and including the Dutch engineering firm RSV; its chief rival was the French firm of Framatome. In the early months of 1975 it appeared that the US–Dutch consortium had the inside track. Feelings in the Netherlands were, however, divided; questions were asked in the Dutch parliament about the advisability of official Dutch government support for a contract with the apartheid regime. When the contract was awarded it went to Framatome. The political debate in the Netherlands may have played a part in deciding the matter in favour of the French; what could not be denied, however, was the generosity of the financial terms offered by Framatome with the backing of the French government. Financing of nuclear exports will be discussed further in Chapters 7 and 8.

Westinghouse, beaten to the future orders for Brazilian nuclear stations by KWU, consoled itself with an order from the Philippines for a nuclear station to be built at Bagac on the island of Bataan. The contract for the Bagac plant was signed in February 1976; but it precipitated another wave of fierce controversy. The Bagac site was only a few tens of miles from an active volcano. Even for an industry which had not baulked at building reactors in earthquake zones this was considered to be going a bit far. The financial arrangements, too, came in for bitter criticism both by opponents of the Marcos regime in the Philippines and by the US Congress; corruption was alleged, while the costs of the project climbed steadily.

While Westinghouse wrestled with the problems of the Philippines plant, it could be thankful to have escaped from the escalating shambles at the Brazilian site at Angra dos Reis. The gradual relaxation of the political climate in Brazil had encouraged senior Brazilian nuclear energy experts to wonder aloud about the desirability of the contract with KWU and more particularly about the goings-on at Angra. Angra 2 and 3, the first two KWU reactors, were to have been located next to the Westinghouse plant Angra 1. But the Angra site had apparently been chosen by the military authorities without consulting any geologists. They might as well have used a map and a pin. The foundation excavations for Angra 2 went deeper and deeper without finding bedrock. By 1982 the site crew had sunk more than 800 steel piles 40 metres long, and the foundations of the Angra 1 reactor, already complete, had begun to tilt towards the gaping hole next door. Brazilians and others doubted whether Angra 2 would ever be completed – or Angra 3 commenced.

As if the troubles in Brazil were not enough to plague the hard-pressed KWU, their other major export contract was with the Shah of Iran. In the late 1970s the first two of a planned programme of some twenty nuclear power stations in Iran were under construction at Bushehr, on the Iranian coast. Then came the fall of the Shah and the advent of the Ayatollah Khomeini. The head of the Iranian Atomic Energy Commission was one of the many senior members of the Iranian establishment to perish in the tumult. Khomeini declared that nuclear energy was the work of the devil, and Iranian Revolutionary officials discussed the possibility of converting the empty concrete containment buildings at Bushehr into grain silos. KWU battled vainly for compensation; by 1983 the status of the Iranian nuclear programme was still chaotic.

In the entire western world, only the French nuclear programme seemed to be proceeding as the authorities had intended. The ferocity of protest, after peaking during the battle of Creys-Malville, had subsided to a sullen murmur; new French stations were ordered in batches, as previous plants came on stream on schedule. The only other country where the nuclear planners were having it so completely their own way was the Soviet Union. The Soviet authorities were so firmly committed to major expansion of the Soviet nuclear programme that they constructed a vast factory to manufacture pressure vessels for the Soviet WWR (PWR) on an assembly line. The factory, called 'Atommash', appears to have suffered some delays and other difficulties; like other Soviet affairs, nuclear matters in the Soviet Union are publicized only when they are successful. The Soviet Union even planted a solid foot in the export market, selling power reactors not only to its client countries, the Comecon block – Bulgaria, Czechoslovakia, East Germany and Poland – but also to Finland.

In the UK the nuclear industry was going through further convulsions. In July 1974 the government had authorized the construction of two nuclear stations based on the British heavy water reactor. But in mid 1976, on the advice of the AEA, the reactor's designers, the government abandoned the heavy water design, which had proved too expensive and difficult to scale up to full power-station size. Despite last-minute hearings by the Parliamentary Select Committee on Science and Technology – long-time heavy-water enthusiasts – the government opted instead for two new advanced gas-cooled reactor stations. These stations were to be ordered, however, not because the electricity suppliers needed the capacity, but because the reactor manufacturers needed the work. UK PWR adherents shook their heads in dismay.

It was not as though the sorry story of the first generation AGRs had by that time had a happy ending. In 1978 the first AGR station, Dungeness B, had not yet started up, thirteen years after it had been ordered; indeed it did not start up until late 1982. But the first reactors at the Hunterston B and Hinkley Point B AGR stations had started up within twelve hours of each other in February 1976. Some said that the teams at the two stations were racing each other. Be that as it may, each station operated for an hour or so, then shut down again for a much more gradual and prolonged period of commissioning. In June 1977, a major cooling water pipe at Hinkley Point B broke; station staff had to rig a fire hose to keep the temperature of the concrete shielding below safe levels. Not to be outdone, in September 1977 station staff at Hunterston B rigged a makeshift pipe in the cooling system and forgot about it. When the reactor was depressurized for maintenance the makeshift pipe delivered several thousand gallons of raw sea-water into the stainless steel core of the reactor. The reactor had to be shut down for many months. All the complex and troublesome interior thermal insulation had to be replaced, at a cost eventually estimated at over £14 million. Capital carrying charges on the shutdown reactor, plus the cost of replacement electricity, pushed the total cost of the oversight well above £50 million. The South of Scotland Electricity Board, to celebrate the advent of Hunterston B, distributed a leaflet to its customers describing how nuclear power was helping to keep down electricity costs. After the Hunterston B fiasco the Electricity Board had to ask for a major tariff increase.

When the UK government nevertheless gave the go-ahead for the construction of twin AGR stations at Heysham B and Torness, in Scotland, GEC – devout believers in the PWR – threw up its hands and asked to be relieved of its management contract for the National Nuclear Corporation, as well as most of its shareholding. The board of NNC was reconstituted yet again; the AEA took over the GEC shares; and the UK nuclear establishment settled down to a further round of quiet civil war.

Elsewhere in the industrial west the battle was not about building gas-cooled reactors or building water-cooled reactors but about building reactors or not building reactors; and not building reactors was becoming the norm. In countries as diverse as Canada, Japan, Italy and the USA, nuclear programmes were slipping inexorably back along the calendar, not primarily because of legal or other opposition but because the electricity suppliers were becoming reluctant to buy reactors. We shall discuss the economic aspects of this situation in Chapter 7. Poor performance and technical problems, too, continued to dog the industry.

In a nuclear plant, you might have expected a light bulb to be safer than a candle. On 28 March 1978, however, a 25-cent light bulb proved to be almost as hazardous as the notorious candle at Brown's Ferry. A technician was replacing the bulb behind a lighted push button on the control panel of the Rancho Seco nuclear plant near Sacramento, California. He dropped the bulb; and it fell into the control panel, shorting out one of the main electrical connections to the reactor's instrumentation. Spurious signals about pressure, temperature, water level and flow, and other important data began flooding into the reactor's main control computer. 'Attempting to match equipment output to the erroneous signals', as the official report put it, the computer shut off the feed-water to the steam generators. The reactor pressure shot up, and the reactor scrammed itself.

The reactor operators watched these goings-on in total bewilderment, 'hampered by the lack of available instrumentation and by equipment responding to the inaccurate signals', as the computer and the reactor itself slowly went berserk. Valves opened and closed by themselves; the reactor pressure and temperature went up and down like a yoyo; gauge readings wandered from one end of the dial to the other. At various stages two of the steam generators went completely dry, with no water in them to take up the heat from the primary cooling circuit. At a later stage auxiliary water filled both main steam generators from bottom to top, so that they began feeding not steam but water into the turbine. By the time the short-circuited power had been restored, seventy minutes after the hapless technician had dropped the light bulb, the cooling system temperature had fallen to below the technical specification limits.

Once the operators had some idea of what was going on they quickly sorted things out. 'The unit remained shut down while data was gathered and sent to Babcock & Wilcox for analysis'; and the analysis concluded that the plant could be returned to service provided that a lookout was kept for loose parts and leaks. The particular sequence of events at Rancho Seco did not, as it happened, involve any safety hazard, although it certainly might have written off some pretty expensive equipment. But it provided a startling demonstration of the fragility of the link between a reactor and its operators, if the link could be reduced to futility by a 25-cent light bulb.

Another factor making utilities increasingly edgy about nuclear situations was the continuing confusion about what to do with spent fuel. The original expectation had been that the operators would discharge the spent fuel from the reactors into cooling ponds at the station; and that within a year the spent fuel would be shipped to a reprocessing plant. The significance of this assumption was not that the operators wanted the fuel reprocessed, merely that they wanted it out of their own cooling ponds. But the near-comprehensive failure of the oxide fuel reprocessing business – only the small French plant at Cap la Hague was in operation and it was fully booked for years ahead – meant that reactor cooling ponds in the USA, Japan, the Federal

Republic of Germany and elsewhere were relentlessly filling with spent fuel. Safety and licensing laws decreed that nuclear power stations were not permitted to continue operation unless they had available space into which to discharge the entire fuel inventory of the reactor in case of emergency. The congested cooling ponds thus meant that more and more stations might soon have to shut down. In the USA a series of proposals and projects were mooted, only to fall one after another into disfavour, either with the federal government or with the governments of the states involved. A so-called Waste Isolation Pilot Plant (WIPP) was planned for a location in the south-western USA; but by 1983 its status was as uncertain as it had been five years earlier. Other plans for various forms of 'away-from-reactor' (AFR) storage facilities ran into similar procedural and other difficulties; and the ponds kept filling.

In the Federal Republic of Germany the Bonn government took a much more orderly approach. They put forward a plan for a massive 'Entsorgungszentrum', incorporating cooling ponds, a reprocessing plant, plants for refining the recovered uranium and plutonium, storage tanks for liquid high-level waste, and a vitrification plant, all located directly atop an underground salt dome, which would serve as the final disposal site for the vitrified high-level waste. The location they chose was the tiny hamlet of Gorleben in Lower Saxony, about 5 kilometres from the East German border. But local people did not take kindly to the idea. They petitioned the State Premier, Ernst Albrecht, to commission an independent study of the Gorleben proposal and to hold hearings before giving it the green light. Albrecht, a leading member of the Christian Democrats, was faced with State elections; he was disinclined to bear the political unpopularity for a nuclear plan put forward by the Federal Social Democrats of Helmut Schmidt. Albrecht accordingly acceded to the request from the deputation; and after he had won the State election he duly commissioned some twenty experts from France, Norway, Sweden, the UK and the USA to prepare an independent assessment of the Gorleben proposal. This Gorleben International Review panel delivered a 2000-page report in March 1979; the report was profoundly critical of the official proposal. On 28 March 1979 Albrecht and some 200 other politicians and nuclear industry people convened in Hannover for six days of hearings to consider the independent report, and to cross-examine its authors and another group of nuclear experts who were in favour of the

Gorleben proposal. The Gorleben hearings, bringing together some of the most illustrious nuclear advocates and critics from many parts of the world, under the TV lights and in the presence of key politicians, to discuss an issue of crucial importance for the future of the German nuclear programme, would have been dramatic in any case; but the hearings were to be given an even more dramatic backdrop by events 5000 miles away.

The spring of 1979 had not been a good one for the US nuclear industry. The discovery of cracks in pipework had led to the forced shutdown of many of the country's BWR stations for checking and re-fitting. Hollywood had stepped in, with the release of a motion picture starring Jane Fonda and Jack Lemmon, entitled *The China Syndrome*. The picture centred on the discovery of a design flaw jeopardizing the safety of a nuclear plant, and the efforts of a TV reporter and a nuclear engineer to expose an official cover-up. It was an enjoyable thriller, but nuclear people dismissed it as lurid and far-fetched.

Two weeks after the release of The China Syndrome, *on 28 March 1979 at 4.00 a.m., several feed-water pumps failed. The pumps were in unit 2 of Metropolitan Edison's new nuclear power plant at Three Mile Island, in the Shenandoah River near Harrisburg, Pennsylvania. The events of the next few hours made Three Mile Island the most famous – or notorious – nuclear power plant in the world.*

When the feed-water pumps stopped, the plant's safety system automatically shut down the turboalternator. As the steam production decreased, the temperature and pressure of the primary cooling water rose, and a 'pilot operated relief valve' (PORV) on top of the pressurizer opened. Eight seconds after the first feed-water pumps had tripped, the reactor scrammed itself. But the decay heat from the fission products in the fuel rods continued to deliver more than 150 megawatts of heat into the primary cooling water. When the feed-water pumps failed, three emergency pumps switched on automatically, but both the emergency feed-water lines were blocked by closed valves; no emergency water reached the steam generators. Neither of the two operators in the control room noticed the closed valves; when the first pump tripped, lighted panels had begun to flash; within minutes there were more than a hundred alarms.

As the reactor pressure fell the PORV should have closed automatically; but it did not, although a panel light went off indicating – wrongly – that it had. The PORV stayed open, bleeding off the reactor's precious coolant – unnoticed. The official report in due course reflected 'Had the valve closed as it was designed to do, or if the control room operators had realized that a valve was stuck open and closed the back-up valve to stem the flow of coolant water, or if they had simply left on the plant's high-pressure injection pumps, the accident at Three Mile Island would have remained little more than a minor inconvenience for Met Ed'. Hindsight is a lot easier than foresight.

Two minutes into the accident, with the PORV still open, the reactor pressure fell rapidly, triggering the two high-pressure injection pumps of the emergency core-cooling system. But the operators, seeing that the level of the water in the pressurizer was still rising, switched off one pump and slowed the other to a trickle. Unknown to the operators, the loss of pressure had begun to allow steam bubbles to form in the coolant. The water level in the reactor vessel began to fall. Sometime within the next two hours it fell below the top of the core. The temperature of the exposed fuel rods soared. The zircaloy cladding softened, and began to melt. The glowing zircaloy reacted chemically with the steam, stripping the oxygen from it and leaving hydrogen gas, which began to accumulate in the growing bubble above the exposed core. Soon some of the ceramic fuel pellets may have begun to soften and melt, like liquorice allsorts under a heat lamp; no one yet knows for sure.

The station staff knew nothing of this. All they knew was that at about 4.30 a.m. the control room received a phone call stating that there was more than six feet of water on the floor of the containment building under the reactor. Sump pumps had already drained some of this water – later found to be radioactive, because of the leaking damaged fuel – into an auxiliary building near by.

At about 5.00 a.m. the main cooling pumps began to shake violently. They were pumping not only water but steam; the operators did not recognize the symptoms and shut off the pumps, which thereupon ceased to drive the cooling water through the core. By this time more senior staff of Metropolitan Edison and the reactor builders Babcock & Wilcox, had been contacted. In a conference phone call the B & W site representative at last asked a question which would have saved a lot of headaches had it been asked two hours earlier: was the block valve between the pressurizer and the PORV shut? It was – as soon as he asked – but much too late. What followed, for the next three days, was an almost unbelievable chronicle of confusion, misinformation, contradictory advice and – for the hapless thousands of people living in south-western Pennsylvania – nightmare uncertainty, the fear that an invisible, undetectable horror might engulf them, and that those in charge were blundering around like lumberjacks attempting brain surgery.

At TMI-2, on the afternoon of 29 March at about 1.50 p.m., everyone heard a dull thud. Nobody realized that it had been a hydrogen explosion within the containment. From Friday to Sunday experts from the Nuclear Regulatory Commission tried frantically to calculate the possibility that the hydrogen in the bubble inside the reactor vessel would reach explosive proportions, and to decide what to do about it. While they sweated, the bubble quietly dissipated of its own accord. Conflicting reports of releases of radioactivity and related hazards led the Governor of Pennsylvania, unable to get a consistent answer from the floundering experts, to recommend on 30 March that children and pregnant women within a radius of five miles from TMI should be evacuated; many people further away joined the exodus.

In the days that followed, as the dust settled and the reactor slowly subsided to 'cold shutdown', the recriminations began. President Carter empowered an official commission of investigation, chaired by John Kemeny, President of Dartmouth College. The NRC set up its own Special Inquiry Group, chaired by Mitchell Rogovin, a Washington lawyer. Metropolitan Edison, its parent company General Public Utilities, and the Atomic Industrial Forum took the view that the accident demonstrated the essential safety of nuclear plants. No one had suffered even the slightest injury; what was all the fuss about? The Kemeny hearings, however, made it all too apparent that in the days of the accident virtually everybody involved, from the White House to the control room, had been out of his depth – and that everything from control room design to evacuation plans left a great deal to be desired. The Rogovin report, published in early 1980, was even more disturbing than the Kemeny report; Rogovin's team concluded that TMI-2 had been within sixty minutes of meltdown on the morning of 28 March, and that a catastrophe had been avoided mainly by dumb luck. Rogovin's deadpan narrative novelization of the accident is a hairraising black comedy, which would be hilarious reading if it were fiction. Very few of those swept up in the events of that long week in Pennsylvania came out of them with much credit.

Neither the accident nor even the black comedy were over when TMI-2 reached cold shutdown. Even to get into the reactor building was impossible for many months. The initial stage involved the release of radioactive krypton-85, a procedure to which the battered townsfolk of Middletown and Harrisburg took exception, their trust in expert reassurance permanently dented. The clean-up also involved gathering and decontaminating some 400 000 gallons of loose radioactive water from the premises. Meanwhile, GPU sued the NRC for $ 4 billion – charging that the NRC had not been strict enough in its regulation; the sound of dropping jaws could be heard all through the nuclear community. Not until mid 1981 was it possible even to enter the primary containment: and then only with extreme difficulty, in full protective clothing and briefly. The first remote TV inspection of the ruined core, by means of a tiny camera inserted through a control rod hole in the top of the reactor vessel, was made only in July 1982; and it brought more bad news. As many had long suspected, the entire core was, in the words of one of the scientists, 'a pile of rubble', a chaos of collapsed and partly melted fuel, still of course fiercely radioactive with long-lived fission products, an unappetizing prospect in any terms. Meanwhile the many seals and packings in the plant, intended to keep it leak-tight, were slowly perishing for lack of maintenance and from exposure to the anomalous conditions inside the plant. Even as this book is being written the accident at Three Mile Island is far from over.

TMI-2 shook nuclear policy to its foundations. Its impact outside the USA was almost as drastic as its impact within. The very moment that the feed-water pumps failed – 4.00 a.m. Eastern Standard time 28 March 1979 – was also the moment – 9.00 a.m. European Standard time – at which Professor Carl von Weizsaecker opened the Gorleben hearings in Hannover. No word of the accident reached the Federal Republic until late the following day. But the news when it came stunned politicians, nuclear advocates and critics alike. The week-long 'trek' from the Gorleben site reached Hannover on Saturday 1 April; estimates of the number at the rally that afternoon ranged as high as 140 000 people. But the nuclear opposition had learned their lesson from earlier battles; a vast police presence was proved to be quite un-necessary, as the rally went off without a single violent incident. The politicians took the point. On 16 May 1979 Premier Albrecht announced, in a half-hour live broadcast on national German TV, that he was rejecting the official proposal for a reprocessing plant at Gorleben. In the months that followed, first one site and then another was mooted as an alternative; by mid 1982 several more had been rejected, and none had yet been chosen. Work continued at Gorleben, preparing a site for disposal of radioactive wastes; but the preparation of the waste itself still lacked a plan, much less a plant, and German power station cooling ponds continued to fill.

In some countries, as we have already noted, the control of nuclear politics tended to rely on tear gas and water cannon. In others, however, the ballot box proved to have a crucial role. The Swedish government had already exhibited a willingness to sustain a dialogue with its electorate, even on ostensibly technical questions like energy policy. In the early 1970s some 6000 voluntary energy study groups, supported by government funds, reported back to the government after the oil crisis, to make their opinions known about the course of Swedish energy policy they considered desirable. The Socialist government of Olaf Palme had, however, a long-standing commitment to a fast growing programme of nuclear power; and this found by no means unanimous endorsement among the study groups or the electorate at large. Opposition to Swedish nuclear plans grew progressively more intense; and the leader of the small Centre party, Thorbjorn Faelldin, revealed himself as an outspoken and dogged opponent of the official nuclear policy. Sweden had a general election in September 1976; civil nuclear policy became a key issue in the election campaign; and the result was the defeat of the Palme government, bringing to an end forty years of socialist government in Sweden. Palme himself, in a post-election interview, laid the blame for his defeat squarely on the nuclear issue. Faelldin became Prime Minister, at the head of a three-party coalition; but the other two parties did not share Faelldin's critical view of nuclear power. After some two years of wrangling in cabinet, over the interpretation of legislation demanding demonstration of a 'safe way' to dispose of nuclear wastes, the coalition collapsed. A general election brought back a minority government. But the nuclear issue did not go away.

In the spring of 1979 Liberal Prime Minister Ullsten rejected, once and for all, calls for a referendum on the future of the Swedish nuclear programme, and authorized the loading of fuel into the Ringhals-3 PWR; two days later came the accident at Three Mile Island. Within a week the government revised its position: a referendum would after all be held before any future nuclear stations were authorized. The referendum duly took place a year later, on 23 March 1980. Voters were asked to pick one of three 'lines'. But by this time all three 'lines' had formulated policies calling for the eventual phase-out of nuclear power in Sweden: Line 1 after twenty-five years with completion of stations already on order, Line 2 ditto with utilities nationalized, and Line 3 after ten years with no further work on incomplete stations. The vote split 39 per cent for Line 2 and 38 per cent for Line 3. Was it, as the nuclear interests claimed, a vindication of the nuclear programme? It is hard to see a policy that calls unambiguously for the rundown of the nuclear programme as a vote of confidence.

The French government reacted to the TMI-2 accident with typical Gallic hauteur; a week after the accident the government confirmed the go-ahead for additional nuclear orders. It also commissioned an independent group of high-level academics to report on the prospects for the French nuclear programme. The group duly reported, in August 1980. If the government had expected a rubber stamp for its policies it cannot have been pleased. The academics found that by the end of the 1980s the fall-off in electricity demand would mean a large surplus of costly nuclear capacity on the system. The Socialist François Mitterrand, campaigning for the Presidency, promised to call a halt to the nuclear programme; but after his stunning victory in May 1981 he bowed to the influence of the French nuclear establishment. He did cancel the most controversial project at Plogoff, which had been the subject of bitter local opposition from virtually the entire population. But almost all the remaining new stations would in due course be built. Nuclear advocates outside France looked longingly at the French example, insisting that the French approach was the right way to pursue nuclear power; no shilly-shallying, no hedging, no compromising with protestors – just get on with it, and gain the benefits of cheap nuclear electricity. Then in July 1982 Électricité de France revealed that reduced electricity demand and the high cost of foreign borrowing had led to a deficit of 8 billion francs; only a tariff increase of some 30 per cent would ease the problem. According to EdF, it was their worst financial crisis for thirty years.

In the US the accident at TMI-2 looked like the kiss of death for an industry already reeling. No reactor had been ordered from 1978 onwards; after TMI-2 there seemed little likelihood of even one more order for the foreseeable future. Some said that B & W, the builders of TMI, had sold their last reactor. The industry struggled on with its existing orders; cancellations whittled these, too, away. Even the plants actually in operation and under construction gave problems.

Westinghouse discovered an embarrassing generic problem with its steam generators in the US and Europe; an 'improved design' proved instead liable to tear itself to pieces. Stations were shut down pending major repairs or even complete replacement of the 'improved' D-3 steam generators. Prolonged exposure to intense neutron radiation was producing changes in the metallurgy of reactor pressure vessels, making them more brittle. The shock of adding cold emergency water during a malfunction could break a brittle vessel wide open. From mid 1981 onwards the NRC agonized over the possible necessity of early and permanent shutdown of several first generation stations.

In California the Diablo Canyon station of Pacific Gas & Electric had been the focus of controversy since the 1960s. By 1981 it had been through court cases and hearings and demonstrations; but PG&E had persisted doggedly, even after TMI-2, trying to get the station into service. Construction was complete; PG&E needed only NRC authorization for fuel loading and low-power testing; and on 19 September 1981 they got it. But the objectors were equally tenacious. With the discipline that had marked nuclear confrontations in the USA ever since the Clamshell Alliance showed the way, protestors converged on Diablo Canyon and occupied the site, determined to prevent the loading of fuel into the reactor. PG&E went to court and got an injunction, requiring the protestors to vacate the site. They did so, and fuel loading commenced.

About forty-eight hours later a young engineer on Diablo Canyon noticed something odd. He called his superiors and invited them to inspect certain important blueprints. The Diablo Canyon site was close to the notorious San Andreas earthquake fault – too close for comfort, in the view of some. Key pipework had special seismic supports. But the young engineer had spotted an extraordinary cock-up. The builders had mixed up the blueprints for units 1 and 2; as a result the seismic supports on the completed unit 1 were the wrong way round, strong ones in the place of weak – and weak ones in the place of strong. The ejected protestors wore ironic smiles as the red-faced NRC rescinded its go-ahead, ordered the fuel removed and suspended the operating licence for Diablo Canyon, pending re-evaluation by an independent consultant.

As if this were not embarrassing enough, the NRC in early 1982 were further affronted to find that the consultant nominated by PG&E as 'independent' had already been a PG&E consultant. By 1983 many observers were concluding that Diablo Canyon would never operate.

The NRC found further egg on its face in early 1982. Despite bitter local opposition they prepared to authorize the restart of the undamaged unit 1 at Three Mile Island; then Metropolitan Edison admitted ruefully that the steam generators in TMI-1 were so badly corroded that they would require drastic retubing if not total replacement. The new chairman of US General Electric conceded that GE did not expect a single further reactor order throughout the 1980s, and was putting its reactor sales activities into mothballs.

In the UK the advent of the Thatcher government in May 1979 gave the battered UK nuclear industry new hope. In December 1979 the Secretary of State for Energy declared that the government would like to see a new nuclear station ordered each year for the decade beginning in 1982; and that the CEGB would be invited to apply to build the first UK PWR. In 1980 the CEGB duly applied to build a PWR station at Sizewell B in Suffolk, and asked the National Nuclear Corporation to prepare a design based on the Westinghouse PWR at the Trojan station in Oregon. However, when the design was submitted, several months late, in April 1981, it proved to be much more expensive than the CEGB had anticipated. In the confusion that followed the Prime Minister appointed the AEA chairman, Walter Marshall, to head a task force and sort things out. A new design was prepared and submitted to the CEGB; they in turn submitted it to the Nuclear Inspectorate. After further delays the government announced that a public inquiry would be held, beginning in January 1983. Nuclear opponents, faced with potential expenses running into millions of pounds, were far from happy about the situation; but they began to gather their forces for what might yet be the climactic confrontation over the UK nuclear programme.

At the beginning of the 1970s the controversy surrounding nuclear power had centred upon doubts about radiation and safety. By the beginning of the 1980s, despite the trauma of Three Mile Island, other questions had taken precedence. Was nuclear power really economic? Could civil nuclear activities really be pursued without aggravating the desperate problem of nuclear weapons proliferation? Would the PWR, the world's most popular reactor, get a new lease of life in the UK? Or would the charge of the light water brigade, like that of its namesake in the Crimea, end in glorious failure?

7 · Nuclenomics

Cost-efficiency was not a major criterion in the Manhattan project. What mattered was whether the technology could produce a bomb at all, and then whether it could produce a bomb in time; both of these criteria were of course satisfied. From then on the aim was to achieve, as one American general put it, 'a bigger bang for a buck'. But the early years of nuclear development, in the USA, the Soviet Union, the UK and France, preoccupied almost entirely with military applications, were insulated to a large extent from the cold wind of everyday economics. Even so, those concerned with peaceful applications saw from the outset that basic economic criteria must be met.

One of the first coherent attempts to put numbers to civil nuclear proposals was that of R. V. Moore of Harwell in the UK, who in 1950 drafted a paper entitled 'Natural Uranium Reactors: Economic Factors in Power Production'. Moore correctly identified most of the quantifiable factors which would still be important thirty years later; but he could not have anticipated some of the ensuing complications.

The saleable product from a power station is electricity. The cost of producing it is partly the cost of building the station, and partly the cost of operating it. The cost of building the station – the 'capital' cost – is an investment, and entails interest charges year by year, as the capital value is gradually written down until the end of the station's useful life. Such capital 'carrying charges' can be shared out among the units of electricity generated, as part of the cost per unit. So can the operating costs – for fuel, labour, maintenance, insurance, taxes, etc. In principle, then, it is easy to assess the cost of each unit of electricity (kilowatt-hour) produced by a station, and on this basis to decide whether the station is a good investment, on its own terms and by comparison with other alternatives. In practice, it is anything but straightforward – as we shall see. Accurate estimates of the cost of building the station depend in turn on accurate estimates of the cost of materials and labour; of the time taken to complete the station and bring it on stream; and of the long-term behaviour of the rates of interest to be paid on the capital involved. The estimated cost of building the station should also take into account the estimated cost of eventual decommissioning and dismantling. Accurate estimates of running costs depend in turn on accurate estimates of the cost of raw uranium; of enrichment if necessary; of fuel fabrication; of reprocessing or other means of management of spent fuel; of final disposal of high-level and other radioactive wastes; and of the specialized security measures required for the protection of fissile materials. The possible income from the station will depend on the eventual price and sale-ability of the electricity produced – a decade and more hence; and on the station's performance – in particular how its actual performance compares to the design specifications used as a basis for the original investment analysis. All of the above listed estimated quantities have proved difficult to get right, and spectacularly easy to get wrong.

For a variety of reasons a station does not operate at its maximum rated design output all the time. Accurate costing must allow for its actual performance, and how well this performance approximates to the ideal embodied in the design. A common measure of performance is the station's 'load factor' – the fraction of maximum possible operation it actually achieves during a certain period. This is sometimes stated as an 'availability factor', or as a 'capacity factor'. 'Availability' is the fraction of time that a power station is 'available' – that is, able to operate. It does not, however, identify periods when the station is operable but is for one reason or another supplying less than its rated maximum output. 'Availability' is not measured consistently. If a station is shut down for refuelling, the availability may be given as a fraction of the entire year. Or it may be given as a fraction only of that part of the year not occupied by refuelling; this gives a higher figure for its availability.

A more valuable measure of performance is the 'capacity factor': the total number of units sent out, as a fraction of the number which could be sent out if the station were to operate at maximum power for the entire year. The capacity factor takes account of low-power operation as well as of total shutdown. Low-power operation of a station may be a consequence of operating problems, or of restrictions imposed by licensing authorities. It also results from the operational characteristics of the whole electrical system to which the station belongs. An electrical utility like the UK's Central Electricity Generating Board operates stations in what is called a 'merit order'. Those stations with lowest total generating costs operate all the time at maximum output, supplying the electricity which is needed all day long and all year long. Such stations are called 'baseload stations'. When the demand for electricity exceeds the baseload – for example on cold days, and at meal-times – additional stations are added to the system to cope with the peaks; they are called 'peaking stations'. The order in which stations are added, which corresponds roughly to the order in which their running costs increase, is called the 'merit order'.

In general, stations with high capital cost but low operating costs are used as baseload stations, for capital charges have to be paid whether or not the stations are in use. Nuclear stations, with their enormous capital cost, are at present invariably used as baseload stations. Accordingly, nuclear stations are expected to operate at maximum output continuously. However, if electrical supply systems build more nuclear capacity, and as stations age, there will come a time when older nuclear stations drop lower in the merit order, and operate below maximum output at least some of the time. The resulting drop in capacity factor will make capital carrying charges of nuclear stations increasingly burdensome in later years; but the precise effect is difficult to anticipate. The cost of a power station of course depends on its size – the larger the more expensive. However, the cost does not increase in proportion; doubling capacity does not double the cost. The relevant measure is the cost per kilowatt of capacity; larger stations cost less per kilowatt – with certain provisos, as we shall see.

Throughout the 1950s and much of the 1960s the economic status of nuclear power depended entirely on the current prices of alternative fuels – coal, oil and natural gas – in national economies. In the USA, as indicated in Chapter 4 (p. 116), the abundance of indigenous oil and gas, as well as coal, kept prices low; accordingly the 'break even' price – at which nuclear and fossil-fuel electricity cost the same – was low enough to make the nuclear option economically doubtful. In Europe – particularly in the UK and France – such was not the case. Coal was available but not, it seemed, in enough quantity to meet the anticipated rapid upsurge in demand for electricity. Nonetheless it was apparent that any move to nuclear power would have to be gradual, and that its cost would not undercut that of coal for some years. Furthermore, it was also apparent that at least a few uneconomic stations would have to be built to give an adequate foundation for eventual economic nuclear power. Both the UK and France found a convenient compromise, building Calder Hall, Chapelcross, and

Marcoule G-1, G-2 and G-3, for the military purposes of plutonium production, and the civil purpose of electricity generation. In the USA, where uneconomic reactors seemed even more inevitable at current fossil-fuel prices, the AEC underwrote the construction of the entire first generation of US civil reactors – eight of which duly proved so uneconomic (or inoperable) that they were shut down permanently by 1970.

Then, in December 1963, Jersey Central Power & Light ordered the 640-MWe Oyster Creek BWR from General Electric, with no AEC subsidy; the time for economic nuclear power in the USA seemed at last to have arrived. However, the cost figures stated at the time proved in due course to involve a subsidy not from the AEC but from General Electric, who clearly regarded the sale as a 'loss leader' to induce further purchases and, presumably, to bring down their own costs to a point at which such sales would return a worthwhile profit. The successful Oyster Creek bid from General Electric was for a plant of 515-MWe output capacity, to cost $ 314 per kilowatt – with the added claim that the plant could in fact produce 640 MWe, reducing the cost per kilowatt to an impressively low $ 108. In the following three years, as indicated in Chapter 5 (p. 134) the hitherto reluctant utilities began queueing up to order light water reactors – more than thirty of them, with sizes escalating past the 1000-MWe mark, even including two 1065-MWe reactors for the Tennessee Valley Authority, in the heart of coal-mining territory. It looked as though the overall costs of nuclear stations had at last dropped below those of their competition.

The second UK nuclear power programme was heralded in April 1964, in a White Paper which effectively conceded that the Magnox design had run its course, and that more compact enriched-uranium designs would now have to be considered. Among these were the UK advanced gas-cooled reactor (AGR) and the US light water reactors (PWR and BWR). The government announced on 25 May 1965 that the second nuclear programme would be based on the advanced gas-cooled reactor, and the CEGB in August ordered the Dungeness B station which was to prove such a disaster (see pp. 144–5). The CEGB published a report analysing the financial case for the Dungeness B AGR station, comparing it with the next closest bid (for a General Electric BWR); the Wylfa Magnox station; and the Cottam coal-fired station, situated in the region of lowest coal costs. In view of the fate which was to befall Dungeness B it is poignant to see the

CEGB's cost calculations, to three decimal places. All three stations subsequently failed to achieve their scheduled performance; it is tempting to allude to the best-laid plans of mice and electrical utilities.

By the mid 1970s the nuclear industry, especially in the USA, was watching intently the progress of generating costs, which were climbing on all fronts, fossil-fuel and nuclear alike. The industry referred to a 'learning curve'; it was assumed that the initial availability and capacity factors of a new plant or size of plant would be low, but that as the learning process identified and extirpated the bugs, the performance figures would improve to the long-anticipated design availability of better than 80 per cent and a capacity factor almost as high. In reality, however, they remained stubbornly much lower. In early 1975 the US Nuclear Regulatory Commission announced that forty-two commercial nuclear plants operating in 1974 had an average availability of 68·5 per cent, and an average capacity factor of 57·2 per cent – by any criterion nothing to write home about. By basic economic criteria nuclear power looked to be in difficulty. By the mid 1970s nuclear economics had begun to receive increasingly close scrutiny, not only from worried electrical utilities but also from academic commentators and nuclear opponents.

The first signs of genuine corporate difficulty had already surfaced. The UK had seen the collapse of Atomic Power Constructions in 1969; and the two surviving UK nuclear consortia had undergone a shotgun marriage under government pressure, to emerge as the National Nuclear Corporation, which was, however, slowly starving to death for lack of orders. In the Federal Republic of Germany, AEG-Telefunken was feeling the chill breeze from its nuclear commitments – troublesome turn-key contracts, starting with Würgassen, the first full-scale commercial nuclear power station in the Federal Republic, which was showing not profits but mounting liabilities. In France, Compagnie Générale Électrique (CGE) were summarily informed by the French government that the French programme would henceforth consist exclusively of PWRs and that CGE would get no BWR orders. Fortunately for CGE, it had not at that stage an unduly large financial stake in a French BWR programme. In the USA, General Atomic was about to pull out of the HTGR business.

Another disturbing economic factor had also begun to attract attention. In late 1974 Irvin Bupp of Harvard Business School and Jean-Claude Derian of the Center for Policy Alternatives at MIT drafted an analysis called *Trends in Light Water Reactor Capital Costs in the US: Causes and Consequences*. Their analysis led them to a startling conclusion. The cost per kilowatt of nuclear stations was shown to be increasing, not decreasing – although a decrease was to be expected if the new larger stations were displaying the anticipated economies of scale. On the data available Bupp and Derian pointed out, for instance, that reactors ordered in 1968, expected to cost only $180 per kilowatt, were in fact costing about $430 – well over twice as much. Furthermore, the difference between costs anticipated and costs eventually realized was continuing to widen. Estimates in 1973 suggested that plants entering service in 1982–3 might cost some $700 per kilowatt; but Bupp and Derian insisted that such estimates were impossible to deduce with confidence from the data, and were in fact 'little more than an educated guess'. What was unambiguously clear was that capital costs of large light water reactors showed no signs of stabilizing, and indeed were still climbing at alarming rates, so much so that the lower fuel-cost of nuclear power stations was in danger of being offset by high capital costs. Bupp and Derian developed their economic critique of the US civil nuclear programme in a book given the sardonic title *Light Water: How the Nuclear Dream Dissolved*. It was a significant addition to a small but growing and influential body of nuclear literature: deeply sceptical analyses by avowed supporters of civil nuclear power.

By the mid 1970s the mounting capital costs of nuclear power stations had reopened an issue which had been crucial to nuclear economics for two decades: the comparative costs of coal-fired and nuclear electricity. The rise in petroleum prices, of course, had put oil-fired base-load electricity abruptly out of the reckoning. But coal, so recently the poor relation of the fossil fuels, was experiencing a startling renaissance. Gradually it became evident that coal, far from fading from the energy scene, was hotly competitive with nuclear energy as a source of electricity.

The 'coal-versus-nuclear' controversy in the US came into sharp focus in late 1976 with the publication of a report by the independent Council on Economic Priorities, entitled *Power Plant Performance: Nuclear and Coal Capacity Factors and Economics*, by a young Harvard economist called Charles Komanoff. Komanoff put together a vast data base on operating coal-fired and nuclear plants in the US to the end of 1974, and subjected the data to a searching statistical dissection. Komanoff's findings were illuminating, and disconcerting. A sample including all commercial power reactors in the US of 450 MWe and over, operating from 1968 to 1975, achieved an average capacity factor of 59·3 per cent – far below industry and government projections of 70 to 80 per cent. Nuclear capacity factors were found to decline steadily with increasing unit capacity. Capacity factors appeared to improve with the age of individual stations; but there was no evidence of the 'learning curve': units more recently installed were not operating at higher capacity factors than comparable earlier units at similar stages of maturity. But capacity factors also varied very widely; the highest lifetime capacity factor for a unit was 77·2 per cent, the lowest 14 per cent. Influences on capacity factor included losses due to scheduled outages for refuelling and maintenance two to three times as high as anticipated in industry projections; equipment malfunctions, especially in steam generators and turbines, and fuel failures; and regulatory restrictions mandated by the NRC. Komanoff likewise investigated the performance of the coal-fired units, noting in particular the effect of requirements for control of sulphur emissions from the plants.

Komanoff's conclusions were uncompromising. 'The study projections of nuclear capacity factors significantly below those of coal, based on currently planned unit sizes, eliminate nuclear power's national cost advantage over coal. Coal is competitive with nuclear for new plants in the north-east, and more economic elsewhere, even with the assumed 7·5 per cent nuclear capacity factor improvement ... Overall, postponing commitment to new generating facilities, where this is permitted by reduced load growth, may reduce generating costs. Postponement would also facilitate choice of more reliable plants through further data base accumulations, capacity factor trend identification, and engineering improvement.'

Komanoff's analysis was hotly disputed by nuclear advocates. But US electrical utilities were beginning to demonstrate the essential validity of Komanoff's conclusions, not only by having second thoughts about ordering new nuclear stations, but also by postponing and even cancelling stations already ordered. For his part, Komanoff was able to set up in business as an independent energy consultant, specifically to pursue the development of this issue, adding to the data base year by year and honing and refining his analysis.

Meanwhile the entire context of the issue was receiving a new kind of attention, with the arrival of the concept of an 'energy strategy'. Energy economics was in turmoil, in the aftermath of the petroleum price increase. The industrial world lurched into a deep and stubborn economic recession. Inflation soared into double figures. Fuel and electricity prices rose inexorably; and customers bought much less than the suppliers had expected. Construction times for power stations, especially in the UK and the USA, were ballooning to beyond a decade. Electricity demand forecasters were faced with a near-impossible task – trying to anticipate the level of electricity use so far in the future, when both the cost of producing the electricity and the price at which it could be sold were vague to the point of amorphousness.

That did not, at first, induce the forecasters to appropriate caution. Until the late 1970s government and electricity supply industry forecasters continued to rely on their old rule of thumb, and anticipate an increase in electricity demand of 7 per cent per year. This increase did not however, materialize: on the contrary, in many industrial countries, the demand for electricity actually fell. Several high-level international working parties were set up, to prepare analyses of this baffling state of affairs. Among them were the Workshop on Alternative Energy Strategies (WAES) and the Conservation Commission of the World Energy Conference, each drawn mainly from the top echelons of the fuel and power industries of many countries. The WAES and WEC reports, published in 1977, each concluded that only a major increase in nuclear energy production could prevent grave shortages of electricity supply in coming years. Many other official studies – by the IAEA, the OECD International Energy Agency and Nuclear Energy Agency, the International Institute for Applied Systems Analysis, national governments, and fuel and electricity supply organizations – came to similar conclusions. But their agitated exhortations fell on unresponsive ears; potential investors were shying away more and more from nuclear commitments.

As the official forecasters and planners floundered, independent analysts and commentators began to put forward quite different views of possible future energy developments. These 'alternative' approaches concluded that nuclear energy might play a much more modest role in the overall energy future; indeed some analysts explicitly excluded nuclear energy from their plans, for one reason or another. There is not room within these pages to discuss the details of the controversy over 'energy strategies'; it has already generated a plethora of books, papers, periodicals, conferences and other dialogues, variously amicable and acrimonious, and remains far from resolution (see the Bibliography for a selection of important contributions to the debate). What is undeniably clear is that within the past decade the economic expectations of the world nuclear community have undergone a drastic diminution.

As the reactor vendors watched their domestic markets slide, they began to look ever more hungrily for foreign buyers. By this time, however, a substantial number of western industrial countries had their own national reactor vendors. The US vendors, who had enjoyed almost a clear field in the nuclear export business until the mid 1970s, found that many of their former client countries were now entering the nuclear export business in direct competition with the USA. Far from being able to sell more reactors in, say, France, Japan, or the Federal Republic of Germany, the US vendors found that the French, Japanese and Germans were even moving in on other traditional export markets of the US vendors – especially in the Third World.

As we mentioned in Chapter 6, a report on nuclear exports to the Third World was commissioned by the US Energy Research and Development Administration (ERDA), and published in 1976 as ERDA-52, under the title *LDC Nuclear Power Prospects, 1975–1990: Commercial, Economic, and Security Implications.* It was deeply sceptical about the prospects on all the grounds considered. Indeed it appears to have embarrassed ERDA; copies rapidly became very difficult to obtain, and it went out of print with suspicious rapidity. Nevertheless the report's scepticism was to be amply vindicated.

In passing, it also spelled out one aspect of the nuclear export business which had hitherto received little attention: the crucial role of nuclear exporting governments and especially their export credit agencies. The conventional view of an export sale is that the exporter is earning valuable foreign currency, and presumably making a useful profit. On closer examination, such is by no means always the case, especially not in connection with nuclear exports. According to ERDA-52 'a very close integration of vendors' willingness to "loss-lead" via costly "turnkey" contracts, generous financial terms from the (US) Export–Import Bank and guaranteed, low-price preferential access to nuclear fuels characterized the first wave of US nuclear power exports to Europe, and a quite similar combination of measures was evident in the initial set of US, Canadian and German exports to LDCs. The German government, for example, underwrote the success of Siemens' HWR sale to Argentina by giving the Argentine government a five-year no-interest loan and subsequent very-low-interest loan, and balance of payments considerations. France managed to sell a reactor unit in Spain in return for loans covering 90 per cent of its cost and agreeing to represent Spanish interests in the Common Market. It is common knowledge in nuclear industry circles that German, US and Canadian vendors "lost their shirts" on their initial sales to Argentina, India and Pakistan.'

But as the US Eximbank began belatedly to tighten its terms in the mid 1970s, the export credit agencies of other nuclear exporters were becoming more accommodating – in particular the French. As of December 1974, for instance, France was offering its overseas customers loans of up to 100 per cent of the cost of the purchase at interest rates as low as 6·3 per cent, repayable over as much as fifteen years – lavish generosity compared to the terms available in the commercial capital market at the time. In general the process would work like this: a nuclear exporter would find an overseas customer; the exporter's government agency would then arrange a low-interest loan to the customer, with a lenient repayment period, to cover most if not all of the cost of the project; and the money would then be paid back to the nuclear exporter in the home country. In effect the entire procedure constituted an oblique but welcome form of subsidy from the domestic government via the overseas customer to the nuclear exporter – a subsidy, that is, from the taxpayers of the exporting government. By the late 1970s such roundabout subsidies were providing a desperately needed lifeline to major nuclear manufacturers in the USA, France, the Federal Republic of Germany, and Canada – and presumably also the Soviet Union.

Even with the help of the export credits, the nuclear exporters were not having matters all their own way. Canadian nuclear people were overjoyed when they beat out international competition to win the first export order for a 600-MWe CANDU power station at Embalse near Cordoba in Argentina. In their eagerness and inexperience, however, they agreed to accept a substantial part of the payment in Argentine currency, overlooking the Argentine inflation rate, several hundred per cent per year. As a result, despite belated and frantic attempts by Atomic Energy of Canada Ltd to renegotiate the financial terms, Canadian taxpayers in effect paid Argentina some C$ 180 million to accept the Embalse CANDU. In June 1978 the Canadian government sacked the president of AECL. Chastened Canadian export negotiators, having won a much more successful contract to build the Wolsung plant in South Korea, were appropriately cautious in their dealings with Romania. Although the Canadians were jubilant when they won agreement to build a CANDU as the first Romanian nuclear plant, with agreement in principle for a good many more, they quickly moderated their rapture when Romania began to show signs of fudging on the financial agreement. By the end of 1982 the status of the Romanian–Canadian contract and even of the first Romanian CANDU remained uncertain; and Canadian nuclear export activities, like Canadian nuclear domestic ordering, had subsided to an ebb sufficiently low to cast doubt on the future of AECL.

As we have already mentioned in Chapter 6, export contracts likewise gave little economic consolation to Kraftwerk Union (KWU). Their jubilation at winning orders for the Brazilian and Iranian nuclear programmes in the mid 1970s turned to dismay and frustration. The problems at the Angra dos Reis site, and the mounting cost of the two reactors ordered, reinforced the chorus of criticism of the entire Brazilian nuclear programme, especially by comparison with the opportunities for future development of Brazilian hydroelectric resources, and even coal. However, the economic confusion surrounding the Brazilian nuclear programme was clarity itself compared to that which engulfed the Iranian programme. Not only KWU, but also the French firm of Framatome and even the UKAEA were caught up in the chaos at the fall of the Shah. Framatome had won orders for the third and fourth Iranian power stations – summarily cancelled by the Ayatollah Khomeini – and the UK nuclear people had been engaged both as consultants and as overseers of quality control at the Bushehr site. Iranian shares in the two French enrichment consortia, Eurodif and Coredif, further complicated the financial tangle. Other export sales were, to be sure, less problematical; contracts with South Korea and Taiwan provided valuable business for US vendors, while work on Framatome's contract for the South African plant at Koeberg continued without difficulty, as did that on the Soviet and Swedish contracts for stations in Finland. But the opportunities for export sales were severely limited, not only because of the ever more intense competition from the various national vendors, but also because of slow – or even negative – growth in electricity demand, and an acute worldwide shortage of capital.

In the UK, government assistance for the beleaguered nuclear industry took the form not of subsidies for exports, but, in effect, of subsidies for domestic orders. As noted earlier, the government had announced in July 1974 authorization for two stations, to use the British heavy water reactor, the SGHWR. By mid 1976, however, the anticipated cost of the full-scale SGHWR had soared to an embarrassing level; the AEA, developers of the SGHWR, recommended its abandonment, and the government concurred. With the rejection of the PWR, this left only the AGR available. It was not, to be sure, needed; the electricity supply industry already had a considerable surplus of energy, and electricity demand was virtually constant. But the power-station construction industry was in crisis: if there were not future orders soon, there was every likelihood that the two major boiler makers and the two major turbine manufacturers might collapse. The government therefore instructed the CEGB to order their planned – but long-delayed – second unit at the Drax B coal-fired station, and a second AGR station at the Heysham site; and the South of Scotland Electricity Board to order a similar twin-reactor AGR station at a green field site called Torness, south east of Edinburgh. The government agreed to compensate the electricity suppliers for the extra costs incurred by thus ordering these stations earlier than they might otherwise have needed to do. The proposal was greeted with disbelief by British PWR advocates, who rehearsed with withering scorn the sorry record of the first-generation AGRs. But the CEGB insisted that the AGRs would in due course prove themselves, and that they were quite happy to order another pair.

On 18 December 1979 the new Conservative government announced that it would like to see one new nuclear station a year ordered in the decade beginning in 1982; and that it would like the CEGB to apply for permission to build the first British PWR. By this time the government and the CEGB had abandoned the argument that nuclear stations would be needed to meet growth in electricity demand; but they insisted that nuclear stations would be needed to replace existing stations as they were retired. The newly constituted Parliamentary Select Committee on Energy held hearings in 1980 on the government's policy statements; and in February 1981 it published a report – coming from a committee known to be strong supporters of civil nuclear energy – which was severely critical of the official policy, of the Department of Energy, and of the CEGB. The Select Committee report declared, among other criticisms, that there was, even in 1981, a strong economic case for cancellation of the two second-generation AGRs. They were also unimpressed by the CEGB's demand forecasting and investment analysis. The official Monopolies and Mergers Commission had been requested by the government to prepare a study of the CEGB. In May 1981 they reported with a yet more devastating comment: 'A large pro gramme of investment in nuclear power stations, which would greatly increase the capital employed for a given level of output, is proposed on the basis of investment appraisals which are seriously defective and liable to mislead. We conclude that the Board's course of conduct in this regard operates against the public interest.'

An independent group called the Committee for the Study of the Economics of Nuclear Electricity was chaired by Sir Kelvin Spencer, Chief Scientist at the Ministry of Power in the 1950s, when the first nuclear programme was announced. It published yet another report in February 1982. Using the official data available, the CSENE concluded that official claims for the comparative cost advantage of nuclear electricity were unfounded. Electricity from existing nuclear plants was more expensive than electricity from coal-fired plants, and was likely to continue to be. The arguments involved assumptions about the accounting procedures, discount rates, the past and future effects of inflation, and the possible future cost of coal. The CEGB and the UKAEA returned immediately to the fray, publishing detailed rebuttals and critiques of the CSENE analysis. When the CEGB published its Statement of Case in favour of the Sizewell B PWR, it put forward economic arguments to demonstrate that it would be cheaper to build Sizewell B and actually shut down coal-fired stations already built and paid for. It was clear that the Sizewell B inquiry would have to address questions not only of amenity and nuclear safety, but also of basic economics – and that the outcome would depend on which set of assumptions appeared most plausible. How a 'planning inquiry' would deal with such convoluted economic controversy was far from clear.

In the US, the economic status of nuclear power was already in question even before the accident at Three Mile Island. The last order for a new nuclear plant was placed in 1978; by 1982 every single nuclear station ordered in the US since 1974 had been either indefinitely deferred or cancelled outright.

The experience of the Washington Public Power Supply System was an egregious example of the pitfalls that had electrical utilities backing away

*from nuclear investments with their hands raised in horror. WPPSS –
pronounced, all too appropriately, 'whoops' – was formed in 1971 by electrical utilities from Washington, Idaho, Montana, Oregon, and Wyoming,
to build a nuclear complex of five 1200-MWe PWRs, on sites at Satsop
and Hanford. The original estimated cost of the complex was $ 3800 million.
By 1975 the costs had passed $ 5000 million. The board of WPPSS issued
bonds, which received a triple-A (top quality) rating. Millions of keen
investors pounced on them.*

*By January 1979 the total debt had reached $ 12 000 million; site prob-
lems were becoming endemic, and murmurs of discontent began to be heard
in the financial world. By 1981, as the plants slipped ever further behind
schedule, costs were increasing at $ 1000 million a month, accompanied by
strikes on the sites and the discovery of yet more structural faults. By late
1981, the estimated total cost of the five units had soared to some $ 100 000
million. In November 1981 WPPSS threw in the towel. Units 4 and 5
were cancelled; the survival even of units 1, 2 and 3 became problematical.
The costs of the cancellations, and of repaying principal and interest on
bonds already issued, threatened to bankrupt some of the eighty-eight
member utilities, and many of the small communities which had so eagerly
climbed on to the WPPSS bandwagon a decade earlier. Many tiny
villages, with less than a thousand inhabitants, now face staggering bills:
and the prospect of paying, over a century or more, hundreds of millions
of dollars for reactors that will never be built.*

In the late 1970s an *ad hoc* group of senior nuclear people from several
countries formed a working party called the International Consultative
Group on Nuclear Energy (ICGNE), and commissioned a number
of working papers. Most of the working papers presented essentially
the received official view of various nuclear issues; but one in particular,
published in 1980, was revealing and disconcerting. Entitled 'The
Viability of the Civil Nuclear Industry', it was written by two young
energy analysts named Lönnroth and Walker, respectively Swedish and
English. Both declared themselves to be firmly in favour of civil nuclear
energy; but their findings gave the industry little to cheer about. They
assessed the total construction capacity of the world's reactor vendors,
and the total plausible future market for new nuclear generating plant
to the end of the century; they found a fourfold excess of construction
capacity. They concluded that the reactor market simply could not
support such an excess: that within the coming decade some reactor

vendors would almost certainly be forced to withdraw from the business. Lönnroth and Walker pursued their analysis into the 1980s, developing it to book length; and events seem to be bearing out their conclusion.

Many official reports and analyses continued to foresee a crucial role for nuclear energy in world development, but they were gravely undermined by a study published in German in 1980, and thereafter in English, entitled *Future Energy Consumption of the Third World*. The study was carried out by a German economist named Markus Fritz, working at the Max Planck Institute in Munich. Its premise was simple. Official analysts insisted that nuclear energy was to play a major role in Third World energy supply: what in fact were Third World countries doing about it? Very little indeed, it emerged. Fritz contacted the responsible energy organizations in 156 countries worldwide, either directly or with the assistance of the Federal German Embassy, and invited the Third World energy organizations to describe their plans. He found that plans as such hardly existed: and that when they did, nuclear came very low on the agenda. Third World countries preferred to rely on their own energy resources, especially hydroelectricity and coal. They did not have grid systems which could accommodate the size of nuclear plant considered economic in industrial countries. They did not have the skilled people to build and operate nuclear plant. Above all they did not have the vast amounts of capital they required for even a modest nuclear programme. The Fritz report made the projections of the IAEA, IIASA and other international agencies look distinctly hollow.

Events in Mexico provided a vivid illustration of the Fritz thesis. In the late 1970s petroleum exploration revealed unexpectedly impressive oil reserves in Mexico; and within short order the country was among the world's leading oil producers. Flushed with the promise of riches from the oil bonanza, the Mexican government announced plans for major industrial expansion – including a programme of nuclear power stations running into double figures. The world's reactor vendors packed their bags and headed for Mexico, each determined to outdo all the others in the generosity of its terms and the lavishness of its hospitality. It did not matter that the one nuclear station in Mexico, the twin-reactor BWR station at Laguna Verde, was already several years behind schedule and losing ground steadily; what mattered was getting the inside track for all those orders to come. Seven different vendors submitted sealed tenders, and sat on the edge of their seats, wondering who would be the lucky winner.

Meanwhile, however, the world oil market moved from shortage into glut; both price and demand sagged. Rumours began to surface that perhaps Mexico was not going to be as wealthy as it had expected. The Canadians, widely tipped as favourites, kept up their courtship; even Prime Minister Pierre Trudeau visited Mexico to woo its nuclear decision-makers. Alas for all the international hopes: on 10 May 1982 the Mexican authorities announced sheepishly that they had changed their minds: they were not after all going to build all those nuclear stations, indeed they were not going to build any more, indefinitely. They returned the sealed bids unopened to the vendors with apologies; and the vendors beat a retreat from Mexico like Napoleon's retreat from Moscow. By August 1982 the world financial community was trying frantically to keep Mexico, the world's largest international borrower, from bankruptcy. The reactor vendors, though licking their wounds, could console themselves with the thought that at least they were only out of pocket a few million dollars; it could so easily have been billions.

8 · Spreading the Risk and Risking the Spread

Plutonium is man-made, an element which essentially did not exist in nature until 1940. Glenn Seaborg and his colleagues at the University of California, using a particle accelerator, first created plutonium, effectively one atom at a time. Seaborg later recalled keeping the world's entire stock of plutonium in a matchbox in his desk. The physical and chemical properties of plutonium were analysed using quantities invisible to the naked eye. Left to itself, a small amount of plutonium undergoes gradual alpha decay, with a half-life of 24 400 years for the commonest isotope, plutonium-239. Curiously enough, when plutonium-239 emits an alpha particle it becomes uranium-235, whose most spectacular property is also exhibited by plutonium-239: the ability to sustain a chain reaction.

Within months of Seaborg's original work it became apparent to the insiders that plutonium-239, like uranium-235, would be potential raw material for a nuclear bomb; and would in some respects be even better than uranium-235. As we have already described, the production of plutonium in quantity was a principal aim of the Manhattan project and of post-war efforts in the USA, the Soviet Union, the UK and subsequently France. However, as well as the fissile property which made it militarily desirable, and the most concentrated source of energy available, plutonium fairly soon showed other attributes of singular unpleasantness. Like radium it proved to be a ferocious radioactive poison, dangerous in quantities of a microgram or less. Plutonium produced in a reactor was – unlike uranium – mainly fissile nuclei, and could without warning reach criticality, either in solid form or in solution. Plutonium workers noted the eerie phenomenon of plutonium oxide 'breathing'. If a tray of the fine fissile dust was filled to criticality, its surface pulsed gently, expanding with the energy release at criticality, thereby going subcritical, collapsing back to criticality, and repeating the cycle endlessly – and of course emitting a burst of neutrons with each lapse into criticality.

What made – and makes – plutonium a matter for urgent concern is not merely its radioactivity, or its fissile nature; other 'actinides' are both more radioactive and more readily fissile. But plutonium now exists in quantities measured in tonnes, is being created daily in ever-larger amounts, and is expected to be a major constituent of the nuclear fuel cycle of the future, not merely as a product but as a raw material. Military facilities, of course, produce plutonium for weapons; civil nuclear systems also produce plutonium in quantity, with as yet no outlet for it. Some governments operated a system whereby they 'bought back' all plutonium created in commercial reactors, crediting the reactor owner with its value against the charges for other fuel cycle services such as enrichment. The US government provided a 'buy-back' service until 1970. But the long-term intention of the civil nuclear community nearly everywhere has been to use plutonium as a fuel. Perhaps 4 per cent plutonium can be added to natural uranium instead of enriching it, for use in thermal reactors; this is called 'plutonium recycle'. As an alternative, much more promising, plutonium can be mixed with uranium in a proportion of perhaps 20 per cent plutonium to 80 per cent uranium for use in fast breeder reactors. But the commercial use of such 'mixed oxide' (MOX) fuel will entail a national and indeed international traffic in plutonium by the tonne – a prospect which seems bound to exacerbate the already intractable problem of restricting access to fissile material.

Such restriction, policing the world's breeding grounds for nuclear weapons, was conceived as a main responsibility of the International Atomic Energy Agency, from its inception in 1956. But its success was then and remained limited. In March 1962 the Agency safeguards system came into being. An incidental difficulty was that it could only be exercised when a national government permitted it to be – and national governments not so inclined were precisely the ones most suspect. It is hard to feel any surge of confidence about the efficacy of the present safeguards arrangements; and in fairness it must be said that this unease is shared by many Agency safeguards staff members.

During the first agreed moratorium on nuclear weapons testing, the Eighteen Nation Disarmament Conference had been set up, by a UN General Assembly resolution of 20 December 1961. Little discernible progress was made at the Conference towards disarmament of those countries already in possession of nuclear weapons and delivery systems. But it would clearly be preferable if, at the very least, countries without nuclear weapons could agree not to acquire them. By 1965 the weapons possessors were putting forward proposals to head off 'proliferation' of nuclear weapons. US and Soviet draft treaties on 'non-proliferation' were presented to the Eighteen Nation Disarmament Conference – by this time usually just called the Geneva Conference, and already showing every sign of becoming a permanent institution – and to the UN General Assembly. Reactions among the non-possessors ranged all the way from whole-hearted support to scornful rejection.

In the preceding decade a number of countries had found themselves confronting the decision: to have or not to have. In almost every case the internal domestic debate was heated. Sweden, whose first power reactor at Ågesta had gone critical in July 1963, gave serious thought to building tactical nuclear weapons with accumulating Swedish plutonium. Military leaders and conservative politicians were vigorous advocates of this policy; but the Social Democratic government, after several years of discussion, at last decided against the idea.

Israel determined in 1957, after the Suez debacle with the UK and France, to secure the option of Israeli nuclear weapons. With French help – and aided by the continuing influx of highly qualified immigrants – Israel built a 26-MWt research reactor at Dimona in the Negev desert. The installation was always identified as being for research, but its operations were kept under a veil of military secrecy; and its annual plutonium output of five to seven kilograms could be equated to a bomb a year. In Israel the domestic debate about nuclear weapons was particularly bitter; the apparently perpetual jeopardy of the country among its Arab neighbours made the issue far from theoretical. The question was quite straightforward: would acquiring nuclear weapons make it easier or less easy to guarantee the future of Israel? No convincing answer was forthcoming either way. But Israel was in no mood to forgo the nuclear option in exchange for promises. Promises could be broken.

From the 1950s onward India developed advanced nuclear technology worthy of comparison with any. India's first research reactor at Trombay went critical in 1956; apart from the nuclear weapons nations only Canada, Norway and Belgium were in the field sooner. In 1960 the 40-MWt CIRUS reactor at Trombay, a joint enterprise between India and Canada, went critical. Canada worked in close partnership with India in nuclear matters throughout the 1950s and 1960s, although the heavy water for the CIRUS reactor, and also the first Indian nuclear power station at Tarapur, were supplied by the USA. India also acquired a plutonium separation plant. Over a period of years, India emphasized her interest in the use of nuclear explosions for civil engineering purposes. As a 'non-aligned' country India endeavoured to maintain her diplomatic distance from both nuclear weapons camps. When in 1964 China, one of India's own most persistent foes, weighed in with a bomb, India made clear her intention of reserving the nuclear option, however vigorous the pious Indian denunciations of nuclear testing and of brow-beating by the original nuclear weapons nations.

On 12 June 1968, the UN General Assembly commended the joint US–Soviet draft Non-Proliferation Treaty. 'Commending', of course, was only an acknowledgement that such a document existed; it needed to be signed and ratified by interested national governments before any of its provisions could be regarded as having more than philosophical import. Article I of the Treaty prohibits the transfer of nuclear weapons (or other nuclear explosive devices, which might as well be weapons) to any states under any circumstances. Article II prohibits Treaty members from manufacturing or acquiring nuclear weapons – but not from preparations up to the point where it is only necessary to fit a weapon together. Article III obliges non-possessors of weapons to accept International Atomic Energy Agency safeguards on all their nuclear activities, to ensure that they do not covertly 'divert' fissile material to nuclear explosives; no Treaty member may supply fissile material to a non-member unless the non-member agrees to the International Atomic Energy Agency safeguards. Article IV says that all Treaty members may, nonetheless, do anything else they wish to do with nuclear energy for peaceful purposes, and may help each other to this end. Article V says that possessors of weapons must agree to provide nuclear explosives for peaceful purposes, under international control and for appropriate charges, to non-possessors who want them. Article VI exhorts members to keep on trying to dispense with nuclear weapons – to find 'effective measures for nuclear disarmament'. Article VII says that members may agree on 'nuclear-free' zones. Article VIII says that a conference to review the Treaty shall be held five years after it comes into force; the first review Conference was held in Geneva in 1975 and the second in 1980. Article IX permits additional nations to become parties to the Treaty once it comes into force. Article X permits a member to withdraw from Treaty obligations on three months' notice if the member decides that extraordinary events, related to the subject matter of the Treaty, have jeopardized the supreme interests of the member-country – or in less fancy language, you may withdraw in three months if you want to.

Taking Article X together with Article II, and noting that final assembly of a nuclear weapon need take much less than three months, if you have the parts ready and are in a hurry, the strictures of the Treaty cannot be called onerous. Yet about one quarter of the nations of the earth have not ratified it. Non-signatories include of course France and China, Argentina, Brazil, Chile, Cuba, India, Israel, North Korea, Pakistan, Saudi Arabia, South Africa, Spain, Tanzania, Vietnam and Zambia.

On 18 May 1974, in the Rajasthan desert in the west of the country, India detonated a 15-kiloton underground nuclear explosion. India thereby became the sixth country to possess nuclear weapons technology, although the Indian spokesmen insisted stoutly that their explosion was for 'peaceful purposes' only. Be that as it may, the Indian explosion underlined with dramatic abruptness the question which had begun to preoccupy many nuclear observers. In the burgeoning world-wide enthusiasm for nuclear energy, it had for some years seemed possible to draw a degree of distinction between civil nuclear systems and their military implications. As of 18 May 1974, such a distinction became very difficult to discern.

During the 1940s and 1950s there had been an aura of awesome mystery about nuclear weapons, about the technology, about its impact on policy, about the unthinkable consequences should such weapons be used. But somehow, by the late 1960s, the growing nuclear arsenals were taken for granted, no longer a matter for conspicuous public concern. Millions of people had been or were still employed in constructing and tending nuclear and thermonuclear bombs – and many thousands of them shared the detailed knowledge which only a decade before had been the most closely cherished secrets. What had been a vocation had become a job like any other. As the global inventory of fissile materials soared, the quality of the supervision dwindled.

The Nuclear Materials and Equipment Corporation reported to the AEC in 1965 that over some six years' activity at its fuel-fabrication plant in Apollo, Pennsylvania, it had somehow mislaid more than 60 kilograms of highly enriched uranium – enough weapons-grade material to make several fission bombs. This material unaccounted for – MUF, in industry jargon – might have been simply the cumulative traces not recovered from scrap. On the other hand, it might not. The AEC forthwith set up a new Office of Safeguards and Materials Management, charged with tightening up controls on fissile materials.

When the first Chinese fission bomb went off in 1964, and proved to be made not of plutonium but of enriched uranium, dumbfounded Western nuclear observers at first assumed it had been stolen – perhaps from Apollo? Satellite photographs of the Chinese gaseous diffusion plant subsequently obviated the need to assume Chinese theft of uranium-235. But the uranium was still missing, although some was later found in scrap; some of it has never been recovered. If it could be imagined that a national government might steal fissile material, what other grim possibilities might exist? On 27 October 1970 the police in Orlando, Florida, received an anonymous message informing them that the sender had a hydrogen bomb, and warning that it would be set off unless one million dollars were paid. The following day another message included a sketch diagram of the bomb – and distraught officials confirmed that it looked all too genuine. There did not seem to be any theft of the necessary nuclear material recorded, but it was also impossible to be sure that no such theft had occurred. Then, to the bottomless relief of the local authorities, a police trap snared the letter-writer: who turned out to be a fourteen-year-old boy, and whose 'bomb' was a hoax. But it might not have been. In the ensuing decade such nuclear hoaxes became – the term itself is alarming – routine.

In the summer of 1971 Kansas State University played host to a deadly serious conference whose theme was 'Preventing Nuclear Theft'; it was apparent that doing so would not be easy. In September 1972 the annual international Pugwash conference of leading scientists from many countries, meeting in Oxford, England, endorsed a statement including the following warning:

The enormous world-wide spread of nuclear fissile material (mainly plutonium) and of nuclear know-how, which is going to occur in the next one or two decades to satisfy the energy demands of the world, constitutes a problem of staggering proportions. It is clear that the management of this problem will necessitate a high degree of international collaboration, if disasters of major proportions are to be avoided. It is difficult to imagine that such an amount of collaboration will be possible unless detente and disarmament make substantial progress in the immediate future. There is a danger, to some degree already present, that processed fissile materials in storage or in transit may fall into the hands of irresponsible, possibly criminal or fanatical groups. The need for ensuring the physical protection of fissile materials, by both international and national means, must be strongly emphasized.

That there were such 'irresponsible, criminal or fanatical groups' at large was amply evident; it quickly became equally evident that they had not overlooked the possibility of nuclear malevolence. Indeed, the notion of nuclear sabotage and terrorism became a popular cliché, amid a spate of paperback thrillers, films and TV programmes on the subject.

Clearly, the impulse towards nuclear malfeasance on every scale is abundantly present. What of the opportunities? Material like plutonium and highly enriched uranium from which nuclear weapons can be made is called 'special nuclear material', 'special fissile material', 'strategic material' or just 'SNM'. The April 1969 meeting of the Institute of Nuclear Materials Management heard Sam Edlow, a consultant on nuclear materials transport, relating a series of experiences, including some of his own. Strategic material in amounts sufficient for dozens of bombs was, according to Edlow, routinely lost, misrouted and overlooked by airlines, trucking companies and freight terminals. A shipment of his, 33 kilograms of 90 per cent enriched uranium travelling from New York to Frankfurt, was mistakenly offloaded at London Airport and left there unattended until the shippers asked the airline about it. A US domestic shipment from Ohio arrived in St Louis with one of three containers of strategic materials – gross weight 385 kilograms – missing. Not until nine days later did the missing container finally turn up – in Boston under a load of shoes.

By 1972 a substantial number of people within the nuclear community were becoming openly worried about the increasingly casual attitude towards strategic material. One of the most outspoken was a nuclear physicist named Ted Taylor, who during the 1950s had been the AEC's star designer of fission bombs at Los Alamos. Taylor had been a contributor to the Kansas State Symposium, as had Mason Willrich, a lawyer and one-time staff member of the US Arms Control and Disarmament Agency. In 1972 Taylor and Willrich were commissioned by the Ford Foundation's Energy Policy Project to prepare a study of nuclear theft. Their work occupied more than a year, during which Taylor visited many of the facilities in the USA, government and private, which shared responsibilities for the secure handling, transport and storage of special nuclear material. With him much of the time was a writer named John McPhee, who chronicled their colloquy in a remarkable book called *The Curve of Binding Energy*, first published as a three-part series in the *New Yorker* magazine in December 1973. The book could have been called 'alarmist' if there had not been at the same time official reports drawing conclusions in broad factual agreement with McPhee's presentation. On 7 November 1973 the US General Accounting Office published its Report to the Congress, *Improvements Needed in the Program for the Protection of Special Nuclear Material*, which was a low-key title for a hair-raising document. The investigation looked into three out of ninety-five organizations licensed to possess strategic material in quantities sufficient to require compliance with AEC requirements for protection. Two of the sample of three failed signally to fulfil the requirements. Facilities had weak physical security barriers, ineffective guard patrols, ineffective alarm systems, inadequate automatic detection devices, and no action plan to deal with a theft of nuclear material. Investigators found that they could enter facilities undetected, climb over fences, pull fencing apart, cut through steel-panelled storage buildings with tin snips in minutes, reach access windows unobserved and almost unimpeded, and in general virtually help themselves as they wished.

What could they do with strategic material once they had it? It had long been dogma that both the necessary knowledge and the necessary technology were beyond the means of any but highly organized efforts – nothing short of a national government, and then only with a major national effort.

It was further assumed that plutonium would have to be made expressly for weapons use, that only 'weapons-grade plutonium' would produce an explosion. The reason given was that different isotopes of plutonium behave differently in a bomb. Plutonium-239 has a very low probability for spontaneous fission, and a high one for neutron-induced fission. But if plutonium-239 is left in a reactor some plutonium nuclei may absorb neutrons without undergoing fission, and become plutonium-240, and then plutonium-241 and -242. Plutonium-240 has a substantial probability for spontaneous fission. Accordingly, a sample of plutonium containing a sizeable fraction of plutonium-240 always contains a significant crossfire of neutrons from spontaneous fission of the -240. Conventional wisdom held that these neutrons would make a bomb made of such material go off prematurely, blowing itself apart before a comprehensive chain reaction could have a chance to take place. Since the higher isotopes of plutonium were considered impossible to separate from plutonium-239 it was felt that plutonium from commercial power reactors would be of little use to prospective bomb-makers.

This comforting thesis began to crumble by the early 1970s. US and European experts came to the conclusion that it might be difficult to predict the performance, the explosive yield of a bomb made of power-reactor plutonium – but that it would very probably go off, with an all too convincing result.

Willrich and Taylor left little doubt about the scale of the consequent problem. Their book, *Nuclear Theft: Risks and Safeguards*, was published in April 1974; it was a landmark study, an instant standard work on a blood-chilling subject. It summarized in dismayingly explicit detail all the information long since available in the open literature, analysed the types of potential bomb-material that the civil nuclear programme would generate, estimated quantities, identified categories of potential nuclear thieves, their motives and modes of operation – nations, political groups, criminal groups, terrorists, fanatics, a who's who of potential nuclear malefactors – and attempted to devise a coherent and feasible programme to repel any attempt at 'diversion' – the delicate industry euphemism which the uncompromising title of the Willrich–Taylor study scorned. Only the part of their book dealing with prevention failed to carry total conviction.

Then, as concern was mounting about the domestic security of fissile materials and nuclear facilities, came 18 May 1974, and the Indian nuclear explosion. Suddenly the nuclear community received a sharp reminder that not only terrorists and criminals might 'divert' fissile material for bombs – national governments, the originators of nuclear weapons, were still very much a major factor in the strategic materials problem. India was the first Third World country to demonstrate nuclear weapons capability; but India was by no means the only possible candidate, Third World or otherwise. The Indian move was far from unexpected; but it was nonetheless profoundly disturbing. For its part India had never concealed its distaste for the Non-Proliferation Treaty. To India, the Treaty represented an attempt on the part of the nuclear weapons powers, especially the USA and the USSR, to calcify the *status quo*, to preserve their enhanced international status while in no way restricting the continuing growth of their own nuclear arsenals or committing them to any significant effort towards disarmament. In the Indian view the threat of nuclear proliferation at the level of national governments was minuscule compared to that posed by the nuclear activities of the major weapons powers. Indians pointed to the amount and availability of nuclear weapons materials in the USA; to the erratic behaviour of US military personnel in Vietnam and elsewhere; to the frequency and variety of criminal or fanatical exploits in the USA; and to the international link-ups between terrorists that might lead to the use of purloined US strategic material in some other part of the world.

Unfortunately, however valid such comments, they emphasize rather than offset the destabilizing effect of increased nuclear activity. The Canadian government was particularly upset by the Indian explosion. The reactor in which India manufactured the plutonium for her bomb was the CIRUS reactor at Trombay, a 40-MWt heavy water research reactor made with the help of Canadian scientists and engineers, during the long and vigorous cooperative programme carried on between India and Canada from the early 1950s onwards. The CIRUS reactor was not subject to International Atomic Energy Authority safeguards; such safeguards had not even been established when it first went critical in July 1960. But bilateral Canadian–Indian agreements led Canada to understand that no Canadian nuclear aid would be used by India to develop nuclear weapons. After the 18 May explosion the Indians simply declared that the device was not a weapon, but a peaceful explosive. The semantic discussion did not impress the angry Canadians, who forthwith cut off all further nuclear assistance to India. In due course other assistance, from sources with fewer scruples or at any rate a more flexible policy, was forthcoming, from France among others. India for her part almost immediately announced a bilateral arrangement to exchange nuclear know-how with Argentina, another Canadian customer, already building a CANDU power reactor at Embalse, and in the market for another.

US nuclear interests took a sanctimonious line that further infuriated the Canadians. US spokesmen referred repeatedly to the 'inadequate' safeguards on the CIRUS reactor – so unlike the scrupulous safeguards imposed on exported US reactors, they implied. Quite apart from the dangers of throwing stones in glasshouses, the US spokesmen overlooked a salient point: the heavy water inventory of the CIRUS reactor was supplied by the USA, whose complicity was accordingly immediate at whatever level of condemnation might be applicable.

The Indian explosion presaged a resurgence of concern about the potential military implications of civil nuclear export activities. The first NPT review conference in Geneva in May 1975 revealed the widening schism between the original nuclear countries and Third World countries about the interpretation of the obligations of parties to the NPT. The suggestion that some kinds of nuclear support should be unilaterally withheld from some prospective clients was greeted with rage and fury by the prospective clients in question. Third World countries declared that it ill became the nuclear weapons powers to express concern about possible weapons developments in other countries – not least because the existing weapons powers had shown not the slightest evidence of compliance with article VI of the NPT, calling for a serious attempt to achieve disarmament. Only last-minute efforts by the conference chairperson, Inga Thorsson of Sweden, kept the review conference from breaking up without an agreed final statement.

Be that as it may, the nuclear exporting countries were already taking steps to bring matters back under control. In conditions of great secrecy they convened a series of meetings of what became known as the Nuclear Suppliers Group; it was also called the London Suppliers Group, after the venue of the top-secret meetings. Participants included the seven leading nuclear exporters – the USA, the USSR, the UK, Canada, France, the Federal Republic of Germany and Japan – who were eventually joined by about eight other industrial nuclear countries. The deliberations of the Group were reported to be directed towards establishing ground rules so that nuclear exporters did not compete for international contracts by offering less stringent safeguards requirements. The Group was also believed to be considering a catalogue of so-called 'sensitive technologies', called the 'Zangger list', itemizing the varieties of hardware – nuclear and other – which might be of use in a nuclear weapons programme: not only entire installations like enrichment plants, reprocessing plants, and heavy water plants, but also components – valves, pumps, electronics, the do-it-yourself bomb-maker's comprehensive shopping list.

It must be assumed that the meetings of the London Group had a good deal to say about French agreements to sell complete reprocessing plants to South Korea and Pakistan, likewise the West German agreement to sell an enrichment plant and a reprocessing plant to Brazil. The public manifestations of this mounting secret controversy included the Flowers and Fox reports of September and October 1976, and the Keeny report, also known as the Ford-MITRE report of spring 1977. The first top-level revelation of the depth of the disagreement came with President Ford's statement of October 1976, in which he declared that henceforth reprocessing and the commercial use of separated plutonium would no longer be regarded officially as essential constituents of the US civil nuclear programme. President Carter, on 7 April 1977, was even more outspoken. Contracts between US nuclear suppliers and overseas customers specified that such customers must obtain permission from the USA before reprocessing fuel supplied by the USA, or even shipping such fuel to a third country with a view to having it reprocessed. Until 1977, the US government had always supplied the relevant permission, on a so-called 'MB-10' form, more or less automatically on request. Suddenly, however, the Carter government let it be known that such MB-10 permission would no longer be granted automatically: and that it might indeed be withheld, in the interests of limiting the spread of separated plutonium.

The particular targets of this shift in US policy included Japan and the UK, and their plan to build a large new reprocessing plant at Windscale, partly financed by the Japanese utilities and intended to reprocess some 3000 tonnes of Japanese spent fuel (see pp. 160–62). The Japanese had just completed their own prototype reprocessing plant at Tokai Mura; amid their bitter protests they were forced to bow to US insistence that the Tokai Mura plant operate only conditionally, for two years, and reprocess only 99 tonnes of fuel. The UK government simply rejected US objections outright, in giving the go-ahead for construction of the THORP installation at Windscale. The Australian and Canadian governments were meanwhile screwing the international tension several notches tighter. Australia, while debating the development of vast new uranium ore bodies, particularly the Ranger proposal (see p. 160), made it clear that they would expect prospective customers to accept very stringent safeguards conditions on any uranium supplied by Australia. The Canadian government went a step – a most unpopular step – further. They notified their customers in the EEC that Canada would require agreement to stringent safeguards on Canadian uranium and what became of it; and when the EEC countries demurred, Canada suspended uranium shipments to Europe. The EEC countries, with sizeable stockpiles of uranium, found the Canadian move more annoying than inconvenient. The USA, however, upset its European nuclear customers even more by banning the shipment of highly enriched uranium to fuel European research reactors, particularly in the Federal Republic of Germany.

The international nuclear situation was clearly deteriorating with alarming rapidity. In May 1977 the leaders of seven major western industrial countries convened in London for what became known as the Downing Street Summit. The seething dissension over nuclear affairs figured prominently on the agenda. Following the summit meeting it was announced that the leaders had agreed to set up an 'International Nuclear Fuel Cycle Evaluation' (INFCE). Its aim would be to study the relationship between civil nuclear activities and nuclear weapons, in the hope of identifying a combination of preferred technologies which would minimize the threat of the proliferation of nuclear weapons capability. INFCE was formally launched at a conference later that year; and in due course involved participants from some sixty different countries and five international organizations.

It was billed as the most extensive and detailed study of civil nuclear policy ever carried out. However, despite its origins in an acute top-level political confrontation, INFCE was thereafter described as a 'purely scientific' investigation which was to refrain from any comment on the 'political' implications of nuclear activities. This disclaimer carried little conviction; the controversy at the heart of INFCE was plainly the bitter dispute between the US and the rest of the world's nuclear community about the commercial use of separated plutonium. But INFCE rapidly widened its focus to cover every aspect of nuclear affairs – notably, and unexpectedly, the unease among nuclear customers about the possibility of unilateral interruption of supplies of uranium, enrichment or other essential materials and services. INFCE participants set up eight working groups: (1) fuel and heavy water availability; (2) enrichment availability; (3) assurances of long-term supply of technology, fuel and heavy water and services in the interest of national needs consistent with non-proliferation; (4) reprocessing, plutonium handling, recycle; (5) fast breeders; (6) spent fuel management; (7) waste management and disposal; and (8) advanced fuel cycle and reactor concepts. Their deliberations were by no means as 'non-political' as their terms of reference claimed; indeed achievement of agreement was sufficiently difficult that INFCE's final report was more than a year late in appearing.

When it was in due course published, in March 1980, its nine volumes proved to be based essentially on the lowest common denominator of policy espoused by the various factions in the world nuclear community. Since the participants in the study had been drawn more or less exclusively from the nuclear community itself, this was not perhaps surprising. The report declared that 'proliferation is primarily a political, not a technical matter', and that there was no magical proliferation-free nuclear technology: all combinations of civil nuclear technologies created similar risks of proliferation. The report concluded on this basis that everyone could therefore go ahead and do what they had first thought of: reprocess spent fuel, separate plutonium, use it in thermal or fast reactors, develop various technologies for enrichment and the production of heavy water, and pursue advanced cycles – all, to be sure, under the watchful eye of the IAEA and subject to its safeguards.

For the amount of effort put into it, the outcome of INFCE must be accounted, especially by nuclear criteria, a damp squib. One un-ambiguous consequence of INFCE was the clear-cut message to the US government from its overseas nuclear partners and clients that they would take profound exception to US arm-twisting in nuclear policies. But even as within INFCE the US government was edging quietly away from its uncompromising 1977 position on reprocessing and the use of separated plutonium, the US Congress was drafting and eventually passing the US Nuclear Non-Proliferation Act, laying down strict controls over US nuclear exports, and making it illegal for the US to supply nuclear technology or materials to countries which did not comply with these controls.

The NNPA, too, at once created embarrassment for the US govern-ment. The USA had been a major supplier of military hardware to Pakistan, despite the long-time insistence of successive Pakistani leaders that the country reserved the right to acquire nuclear weapons if they were adjudged necessary. In December 1978 evidence emerged that Pakistan's earlier efforts to build a reprocessing plant – with the increas-ingly reluctant help of the French – were in fact a remarkable ruse to conceal Pakistan's actual intentions. It was revealed that a Pakistani who had worked briefly at the Almelo gas-centrifuge enrichment plant of URENCO (see p. 81) had acquired highly classified technical information, enabling Pakistan to commence construction of a top-secret enrichment plant at a site called Kahuta. Specialized components for the plant were purchased through front-companies in western Europe. Diplomats and a journalist who attempted to learn more about the Kahuta project were attacked and beaten by professional thugs. The developments led some observers to believe that Pakistan was being backed by Libya, to develop an 'Islamic' bomb.

Elsewhere on the sub-continent, India too found itself up against the US Nuclear Non-Proliferation Act. When the USA supplied India with its first nuclear power station, at Tarapur, a twin-reactor replica of Dresden 1, the deal included a lifetime supply of fuel for the Tarapur plant. After the Indian explosion, however, the USA acknowledged belatedly that it had supplied the heavy water for the CIRUS research reactor used to make the plutonium for the 'device'. According to the provisions of the NNPA, unless India agreed to comply with the tightened safeguards now required by the USA, the USA could not legally supply any more reload fuel for Tarapur. India responded with ire, declaring that if the USA were to abrogate its contract India would no longer be bound by the contractual provisions about not reprocessing existing spent fuel from Tarapur without US permission. The issue poisoned the air between the USA and India for many months, although the US Congress voted narrowly – after intense lobbying by the Carter administration – to allow two shipments of fresh fuel to Tarapur. In mid 1982, in one of its more idiosyncratic foreign policy moves, the Reagan administration announced with pride that it had resolved the issue; fresh fuel for Tarapur would be supplied by the French, thus satisfying both the US NNPA and the Indian desire to get Tarapur back on stream. Whether this had anything to do with the original objectives of the NNPA did not apparently come into the reckoning.

The NPT itself was becoming almost equally fragile. The 1980 review conference was acrimonious and ineffectual; this time not even last minute heroics could produce an agreed final statement, and the conference broke up without one. The failure of the weapons powers, and indeed of virtually every other national government on the planet, to find some way to get to grips with the threat of nuclear annihilation had, however, at last begun to attract dismayed attention, initially in Europe, and then, with startling suddenness, in the USA. The issue is beyond – albeit only just beyond – the framework of this book: but it put the proliferation issue into a new – and long-overdue – public light. The *Bulletin of the Atomic Scientists* ran an essay competition in 1980, for the best entry on 'How to Eliminate the Threat of Nuclear War'. The winning essay proposed, far from frivolously, that the key to nuclear disarmament lay with Third World countries. The weapons powers, said the author, had shown no sign of any genuine desire to achieve nuclear disarmament. Third World countries should therefore demand that the weapons powers agree on certain gradual steps towards disarmament – for instance, a comprehensive test ban, a freeze on existing weapons stockpiles, cessation of the manufacture of new fissile materials, and a timetable for actual retirement of weapons. Failing such steps, the Third World countries should announce that they too intended to join the weapons club: they should use the threat of 'horizontal proliferation' to put a brake on 'vertical proliferation'.

Other commentators went even further. US energy analyst Amory Lovins was senior author of a book entitled *Energy and War*, published in 1980, which took the position that civil and military nuclear activities were inseparable: nothing short of a phase-out of civil nuclear activities worldwide would provide an adequate basis for the banishment of nuclear weapons. The view was reinforced by the publication in late 1981 of a report prepared at the request of the US NRC by Emmanuel Morgan, a former IAEA inspector, on the efficacy of IAEA safe-guards. The Morgan report declared that existing safeguards were incapable of detecting diversion to weapons use of a significant quantity of nuclear material in any state with a moderate to large nuclear establishment; and that even if diversion were suspected, the IAEA would be able to do very little about it. The IAEA took deep exception to these criticisms, but their rebuttals only served to confirm the substance of many of Morgan's comments.

The weapons countries themselves continued to behave in ways guaranteed to excite suspicion and cast doubt on the putative separation of civil from military nuclear activities. In the autumn of 1981 the US government revealed that it was short of plutonium for the new generation of nuclear weapons proposed by President Carter, and enthusiastically endorsed by President Reagan. The US administration suggested that it might be a good idea if the US government were to take the spent fuel from the ponds of US civil nuclear stations, and reprocess it in the US military installations at Savannah River and Hanford, to recover the plutonium for weapons use. The idea horrified the US electrical utilities, who had spent twenty years trying to convince their customers that nuclear electricity had nothing whatever to do with bombs. It was further revealed that the US nuclear weapons laboratories were studying the application of laser isotope separation (see p. 82) to plutonium, to extract purified plutonium-239 from power-reactor plutonium. The Reagan administration also asked the UK government to sell it some 5 tonnes of plutonium from the stockpile of civil plutonium in the UK. When the story broke, in October 1981, the UK nuclear authorities insisted that the 5 tonnes of plutonium were for the proposed Clinch River breeder reactor, and nothing to do with the US weapons build-up. But critics – among them senior CEGB staff and Sir Martin Ryle, the Astronomer Royal – pointed out that provision of plutonium from the UK would free an equivalent amount from the US government's stockpile for use in weapons. In mid 1982, after a fusillade of controversy, the US government withdrew both proposals. But the scepticism about the separability of civil and military nuclear activities was by this time widespread and deep-seated.

Just how deep-seated had been made all too clear on 7 June 1981. The IAEA Board of Governors was holding its annual meeting in Vienna. While the delegates deliberated a squadron of F-16 fighter aircraft and bombers took off from a military airfield in Israel. Flying low across the desert to evade radar detection, the planes swooped down on the Tuwaitha Nuclear Research Centre near Baghdad in Iraq. Within moments bombs and rockets had shattered the 40-MWt Tamuz-1 research reactor, leaving it in ruins. Israel's Prime Minister Menachem Begin, announcing the raid, declared that it had been carried out because Iraq was using the Tuwaitha facility to manufacture nuclear weapons. He added that the timing of the raid was because the reactor was planned to go critical for the first time later in 1981, and that after that time it would be too dangerous to destroy it, lest the fall-out reach Baghdad, 16 kilometres away. This rationalization was patently nonsensical; no conceivable form of attack on the Tamuz reactor could have produced significant fall-out over Baghdad; and many observers concluded that the Israeli raid had more to do with the impending Israeli elections. Begin's prospects had been thought doubtful; but after the drama of the Tamuz raid, Begin won the election handily.

The IAEA, for its part, had no doubt about the real significance of the raid. On 9 June IAEA Secretary-General Sigvard Eklund told the Board of Governors that the Israeli attack was also an attack on the Agency's safeguards. The Tamuz reactor, like all Iraqi nuclear activities, was under these safeguards; the reactor had been visited in January 1981 by two IAEA inspectors, and given a clean bill of health. After the raid IAEA inspectors once again visited the Centre, and verified the presence and condition of all fuel assemblies and other nuclear materials on the site. Iraq had been a party to the Non-Proliferation Treaty since it came into force in 1970. Israel, on the other hand, was not a party to the NPT, and had been operating the 25-MWt Dimona research reactor since 1967 – a top-secret installation under no safeguards of any kind. The day after the Tamuz raid Colonel Gadafy of Libya declared that it was now time for the Arab states to destroy the Dimona reactor. The putative borderline between civil and military nuclear activities was beginning to look more like no-man's-land.

9 · The Nuclear Horizon

Take a deep breath. By 31 December 1981 there were 75 power reactors within the USA licensed to operate, with a capacity of 57 000 MWe. There were 32 power reactors in operation in the UK, with a capacity of 7600 MWe. There were 35 power reactors in operation in the USSR, with a capacity of 14 000 MWe. There were power reactors in operation in Argentina, Belgium, Bulgaria, Canada, Czechoslovakia, Finland, France, the German Democratic Republic, the German Federal Republic, India, Italy, Japan, the Netherlands, Pakistan, South Korea, Spain, Sweden, Switzerland, and Taiwan. In all there were 271 power reactors in operation, with a total capacity of some 153 000 MWe. There were a further 239 power reactors under construction, with a total capacity of 222 000 MWe, in the above countries and also in Brazil, Cuba, Hungary, Mexico, the Philippines, Romania, South Africa and Yugoslavia. Most of the above countries and some others also possess research reactors; some of these, in for example Israel, Norway, and Poland, were comparable in heat output – and plutonium production – to small power reactors.

Many of the above countries had, or planned to acquire, some if not all of the other components of the nuclear fuel cycle: uranium mines and mills, hexafluoride plants, enrichment plants, fuel fabrication plants, fuel reprocessing plants, and waste management facilities. Six countries were known to possess nuclear weapons technology; others were not far behind.

Yet despite this vast industrial base the world's nuclear community in the 1980s is – for the first time since its inception – profoundly apprehensive about the future. The euphoria of the mid 1970s has all but evaporated. There are still, to be sure, pockets of enthusiasm – notably in the UK, France and the Soviet Union; but elsewhere the gloom on nuclear faces has been progressively deepening. In the mid 1970s, projections by the International Atomic Energy Agency, the Organization for Economic Co-operation and Development, the European Economic Community, the US Atomic Energy Commission, the UK Atomic Energy Authority, the French Commissariat à l'Énergie

Atomique, and many other organizations and individuals in the nuclear field asserted that nuclear energy would supply 50 per cent or more of world electricity requirements by the year 2000. Estimates of world-wide nuclear generating capacity in the year 2000 ranged at least as high as 4 500 000 MWe – that is, 4500 power reactors of output as high as the largest now operating. These sky-high visions have long since faded.

It has been a startling and drastic turn-around. From 1970 onwards the nuclear option had begun to play a major role in general energy policy, particularly among the industrialized nations. The desire to reduce dependence on the petroleum-exporting countries, especially those of the Middle East, manifested itself in a determination on the part of the USA, France, Japan and other vulnerable Western countries to expand their nuclear electrical generating capacity as rapidly as possible. The marginal cost advantage of electricity from nuclear stations was dramatically enhanced by the oil price rise, coupled with the renewed insistence of coal-miners in the UK, the USA and else-where on hefty pay rises. Attempts to boost coal output were further hampered by the prolonged economic shakiness of the industry in the UK and France, and a bitter battle about air quality standards and the effects of strip-mining in the USA. Nuclear energy seemed by comparison a manageable technology, ready to expand to meet the prognostications of future demand with minimal effect on the environ-ment and on the health of its workers. Concern about a possible imminent shortage of uranium could be set aside by referring to the forthcoming role of the fast breeder reactor, improving fifty-fold the utilization of uranium and – in a country like the UK – even suggesting that importing uranium would be unnecessary for the indefinite future.

By such time, it was anticipated, other nuclear energy technologies would be playing their part. Nuclear-powered ships, including cargo submarines, tankers and bulk carriers would be able to cruise the oceans at will, with no fear for fuel costs. Nuclear generating stations would be built, not merely on remote coastlines, but in near-urban locations, where the heat output from their turbines, formerly wasted, would supply district heating to entire neighbourhoods and industrial estates, incidentally bringing fuel costs yet lower, and boosting efficiencies to unheard-of heights. Meanwhile new reactor designs, developments of the high-temperature reactor, would take over from the dwindling supplies of fossil fuel, and supply process heat for industry; nuclear steel-making seemed likely to be the first breakthrough on this front. As a demonstration of the ways in which nuclear energy would help to alleviate other resource problems, desalination of sea-water using the low-temperature heat from reactors was expected to become an important contribution to water supplies. Desalination was a service expected to be especially welcome in some developing countries, whose burgeoning energy requirements would be impossible to fulfil from remaining fossil reserves, leaving only the nuclear avenue to development.

Such a view had an undeniable appeal. But it sidestepped a number of knotty questions, to which even in the 1980s answers are not yet fully forthcoming. On a purely economic level the problem of building nuclear capacity at desired rates and of desired magnitudes has proved formidable. Over the last century the trend in energy supply, from wood to coal to oil to gas, has always moved in the direction of inherently simpler technologies and sources. The nuclear option is an abrupt departure from this trend. It is doubtful whether it can be introduced quickly enough both to take over from previous sources, and to sustain envisioned rates of growth. Major programmes all over the world are finding themselves up against shortages of money, of material resources, and of adequately trained manpower.

Other problems persist: recall the hazards arising from the radio-activity of the materials of the nuclear fuel cycle. The effects of low-level radiation, and the gradual build-up of man-made environmental radio-activity, are acutely difficult to ascertain; if, in due course, they are found to be a serious liability to global ecology and to the genetic vigour of living things, it may well be too late to remedy the situation. The same considerations apply, of course, to other man-made contaminants like heavy metals and persistent chemicals; the nuclear worry is not unique except in so far as we know already the genetic vulnerability of organisms to radiation, and do not know how to extrapolate to an entire planet or to many generations of human exposure.

A more acute problem of a similar nature arises from the accumulation of high-level radioactive wastes from fuel reprocessing. Fission products like strontium-90 and caesium-137 remain dangerous for hundreds of years, actinides like plutonium-239 and americium-241 for hundreds of thousands of years. No human artifact can guarantee to isolate such substances for such a period. The volumes of such wastes are, to be sure, small by comparison with wastes from some other energy technologies, pit spoil and ash from coal-fired stations, or sulphurous sludge from stack-gas scrubbers, for instance. If high-level wastes can be solidified, into tough borosilicate glass, their mobility can be minimized. But the questions remain – not so much technical as ethical. Are we justified in availing ourselves of nuclear energy, if by so doing we impose an effectively permanent burden on our descendants? On the other hand, are we justified in confining our energy-requirements to fossil fuels, and thereby depleting them to a pittance for future generations?

More immediate problems also arise. The nuclear fuel cycle abounds with complex technology, in installations which contain awesome amounts of radioactive materials. Have we sufficient assurance that such installations can be guaranteed to operate safely? A breach of containment, either by accident or by sabotage, could release enough radioactivity to render a vast area indefinitely uninhabitable. Some analysts, notably those who drafted the AEC's WASH-1400 Reactor Safety Study, concluded that the probability of a major accident to a light water reactor is minuscule. But their conclusions did not go unchallenged, and within five years they fell under the shadow of Three Mile Island. There remain also the questions of other reactor designs, of other fuel-cycle installations like reprocessing plants, and – unfortunately – of sabotage and even military attack.

The most dismaying problem of all is probably that of fissile material security: guaranteeing that potential nuclear weapons materials – plutonium-239, uranium-233 and uranium-235 – do not fall into the wrong hands, whether of trigger-happy governments or of terrorists. The possibility is far from hypothetical, and does not seem amenable to easy solution. US nuclear interests at one stage proposed the establishment of a National Fissile Material Security Service – a federal agency like the FBI or the CIA, responsible for overt and covert supervision of the nuclear fuel cycle. It is not an attractive idea, indicating a trend towards authoritarian centralized control of a society, of the kind most likely to precipitate just the social unrest which would most readily lead to nuclear upheaval. It also prompts mention of William T. Riley, former top security officer of the AEC, who was sentenced in February 1973 to three years' probation for having borrowed $239 000 from fellow AEC employees, and failing to pay back over $170 000, having used most of the money for betting on horse races. Who, in a Nuclear Security Service, will watch over the custodians?

Fred Iklé, then director of the US Arms Control and Disarmament Agency, speaking in January 1975, pointed out a grimly sobering aspect of the situation. With nuclear technology and materials ever more widespread, what would happen if a nuclear explosion wiped out, say, Washington? The USA might have no firm idea who had done it – and thus the entire monolithic theory of strategic mutual deterrence crumbles into blatant ineffectuality. Curiously enough, at a time when access to resources seems likely to provoke one showdown after another, the widespread dissemination of nuclear technology will undoubtedly have an unexpected, if discomforting, equalizing effect. No matter how small a nation, no matter how small a group of participants, if they have access to nuclear weapons, or even to quantities of long-lived radioactivity, their voice must be heard with respect. From this point of view a country with many vulnerable nuclear installations is offering hostages to all and sundry.

Large power stations – whether nuclear or fossil-fuel – and their related electrical grid systems are vulnerable in a number of ways, both during construction and when in operation. Nuclear power stations cannot readily be built economically below a certain size, at least 100 MWe, and are usually believed to be more economic an order of magnitude larger. Unfortunately stations of such a size take a long time to build – at least five years, and in some cases as many as ten. This means that such a station is being built in a planning vacuum; credible forecasts of electrical demand more than five years ahead can no longer be made. Stations of such a size are prone to engineering problems of many kinds, during construction and during operation; and a small malfunction may make a 1000-MWe station unavailable for many months, at prohibitive cost to the operator. Precisely such a coincidence of effects arose in Britain in the 1960s. On the basis of a forecast 8 per cent growth in electricity demand the Central Electricity Generating Board ordered a vast programme of new capacity, including the ill-fated AGR stations. These stations and the forty-seven very large new turbo-generator sets, of 500 and 660 MWe, which were used in both nuclear and fossil-fuel stations caused seemingly endless problems. Fortunately for the CEGB, the forecast growth in demand was also drastically inaccurate; the actual growth rate was less than 3 per cent, and the unavailable stations were not needed anyway. The losers were the electricity consumers and the taxpayers, who had to finance the unnecessary investment in new capacity, through electricity rates and taxes.

Many commentators have begun to question the alleged economies of scale of very large power stations. Economies may indeed appear on paper; but if construction delays, unforeseen engineering problems, and subsequent breakdowns are allowed for in calculations the economic size of stations seems much smaller than those now being ordered. Furthermore, a power station producing 1000 MW of electricity also produces perhaps 2000 MW of heat – far too much to be usable even in the largest industrial installation. So such giant stations have to be sited at a remote location, where the heat can be discharged to the surroundings as waste – usually in wilderness or amenity areas, which must be further disrupted to provide transmission facilities, which in turn entail further losses of useful energy.

These various problems have seriously undermined earlier nuclear confidence. In August 1982 the IAEA published a supplement to its popular *Bulletin*, celebrating the Agency's twenty-fifth anniversary. The IAEA has always been the world's foremost nuclear cheerleader, taking the rosiest possible view of prospects for future development of civil nuclear energy. But the anniversary supplement included an article by H. J. Laue, Director of the IAEA's Division of Nuclear Power, which was the bleakest assessment ever to emerge from within the Agency. According to Laue:

Even with the oil crisis of the mid 1970s and the various national programmes to switch from oil to other energy sources, the capacity projections made in the late 1970s for 1990 and 2000 plummeted. The projections for 1990 and 2000 published in the 1980 (IAEA) Annual Report were one-third to one-fifth of those published in the 1973–1974 Annual Reports. This decline is in spite of the fact that the price of oil has increased by a factor of at least seven during those six years and that hydrocarbons were widely realized to be in short supply. Although the 1990 capacity projections show signs of 'bottoming out', some recent studies indicate that the actual turnout in 2000 could be as much as 20 per cent lower than the projection reported by the Agency in 1980.

The reasons for this drastic reduction are many.

First of all, there was the economic situation: the more efficient use of energy, the relative decrease of highly energy-intensive production, and the economic recession in industrialized countries led to a much slower increase in demand for electricity. Consequently, new orders for nuclear power plants did not materialize in some countries.

In addition, the introduction of nuclear power on a large scale was curtailed through lack of public confidence, arising from concern about reactor safety and the disposal of nuclear waste, and from unbalanced public perception of various risks. As a result, many countries hesitated to take long-term decisions on nuclear power because of their political implications.

Finally, as the International Nuclear Fuel Cycle Evaluation Study (INFCE) concluded, there is the possibility that civilian nuclear facilities could be misused to produce nuclear weapons (although this is not the usual or most efficient route). This possibility became a major concern of the public and the governments in some supplier countries, and has hindered the further development or introduction of nuclear power in both developed and developing countries . . .

Many nearly complete or operating nuclear power plants have been beset by financial, management, licensing, and technical problems. The number of such plants has grown to around thirty worldwide, corresponding to a total capacity of 27 000 MWe. Of these plants seventeen have never operated or were stopped indefinitely when between 30 and 100 per cent complete: they include Bushehr in Iran and Zwentendorf in Austria. During the first half of 1982 alone, thirteen US plants joined this category, primarily because of mismanagement, financial troubles, and reduced need for electric power; political uncertainty also played a part. Seven plants worldwide have been inoperable for at least two consecutive years for repair, backfitting and licensing reasons, and several are at present limited to less than half-power due to serious steam-generator problems . . .

Thus the primary motives for deciding to introduce nuclear power, reliability and economics, have not received the attention which was necessary to keep nuclear power a viable option.

The final paragraph of Laue's paper concluded:

The future programmes of the IAEA, based on a quarter century's experience in the peaceful use of nuclear power, will address these priority questions, and assist the developing Member States in using this important source of energy more extensively than in the IAEA's first twenty-five years. Unless these challenges are accepted, the viability of nuclear power for the future supply of energy cannot be assured either in industrialized or developing countries.

It was scarcely an encouraging outlook.

Within the past decade, attention has instead been turning to different energy philosophies and technologies. There is of course not space in this book to do justice to the complex and fascinating question of energy options and strategies. A few comments must suffice to indicate key aspects of the argument; see the Bibliography for further material.

The most important change in emphasis is the attention now paid to the ways in which we actually *use* energy. The majority of energy required in an industrial economy is not high-grade or high-temperature energy like electricity but low-grade heat. In some countries, like Japan, Israel, Australia and parts of the southern USA, solar energy has long been used to heat and cool buildings and provide hot-water supplies; interest in low-temperature solar energy applications is now burgeoning worldwide. Even in higher latitudes it is being realized that solar energy can provide a useful input to the total energy supply, even if only for the first stage of water heating. Capital costs are still high, but are expected to come down as the technology matures and acquires more widespread applications. Wind, too, is being recognized as a possible contributor to energy supply in some regions. Like direct solar energy, wind is an inherently decentralized source of energy; both direct solar energy and wind may make an economically significant contribution to energy supply, as the advantages of decentralized sources are recognized.

Geothermal energy – energy from the hot interior of the earth's crust – is also under investigation, as a source both of direct heat and of steam to drive turbogenerators. Its availability will vary from region to region, but may be substantial. Biological processes for generating fuels like methane from organic waste materials have long been established in some localities; as the prices and availabilities of fossil-fuel supplies become less favourable, biogeneration may play a newly economic role. More exotic technologies have also been proposed, including offshore structures to tap the energy of ocean waves, and to utilize the temperature gradients between the deep ocean and its surface.

Advanced conversion technologies have also given a new lease on life to the fuel not long since written off as too dirty and too dangerous: coal. Fluidized bed combustion makes it possible to burn even low-grade high-sulphur coal cleanly, cheaply, and efficiently, while minimizing air quality problems. In due course production of liquid and gaseous fuels from coal also seem likely to be major industrial activities.

Above all, there are untold opportunities to improve the efficiency with which we use all these various kinds of energy. Unglamorous technologies like thermal insulation, heat pumps, and the combined generation of heat and electricity can increase spectacularly the amount of use we get from every unit of energy converted.

Advocates of alternative energy technologies do not claim that any one will provide for all requirements. They say rather that a mix of different sources, matching demand in quality as well as quantity, is feasible and achievable within present constraints of finances, resources and time. They point to the funding of energy research and development, which in the past thirty years has been concentrated on nuclear energy almost to the exclusion of other technologies, even those based on fossil fuels, like coal gasification and liquefaction. If even a modicum of the available research and development effort were redirected into the alternatives, they feel that the nuclear option would soon appear neither the only option nor the best. Others disagree; in their view only the rapid developments of nuclear technology will supply the energy needed by the people of the earth.

We stand today before an abundance of potentials and possibilities; the options are still open. Within the present generation they will almost certainly be foreclosed. The decisions now impending will affect not merely global energy supply and demand, but the entire organization of our global society. We, the people of the world, must be party to these decisions. Before we commit ourselves and our descendants to a nuclear future, it is vital that we concur in and understand the nature of the commitment. If we undertake it now we do so for all time.

Appendix A

Nuclear Jargon

ABCC: Atomic Bomb Casualty Commission; US organization in Japan responsible for victims of Hiroshima and Nagasaki bombs.

ACRS: Advisory Committee on Reactor Safeguards; responsible for assessment of safety of reactors licensed in the USA.

Actinide: one of the heavy elements actinium, thorium, protactinium, uranium, neptunium, plutonium, americium, curium, berkelium and californium, all of which are chemically very similar; actinides of interest are those which are long half-life alpha-emitters.

Activation: absorption of neutrons to make a substance radioactive.

AEA, UKAEA: United Kingdom Atomic Energy Authority.

AEC, USAEC: United States Atomic Energy Commission.

AECL: Atomic Energy of Canada Ltd.

AGR: advanced gas-cooled reactor.

Alpha particle: high-energy helium nucleus (two protons, two neutrons) emitted by some radioactive nuclei.

Atom: see pp. 23–4.

Becquerel (Bq): unit of radioactivity; one radioactive decay per second.

Beta particle: high-energy electron emitted by radioactive nucleus.

Boron: powerful absorber of neutrons used – usually in alloy steel – for reactor control rods, etc.

Breeder: reactor which produces more fissile nuclei than it consumes.

Breeding gain: proportional increase in fissile nuclei in fuel after its removal from breeder.

Burner: reactor which consumes more fissile nuclei than it produces.

Burn-up: cumulative output of heat from reactor fuel; directly correlated with build-up of fission products; usually measured in megawatt-days per tonne of uranium.

Butex: organic solvent used in reprocessing irradiated reactor fuel.

BWR: boiling water reactor.

Caesium: particularly caesium-137; fission product, biologically hazardous beta-emitter.

Calandria: in pressure-tube reactor designs, tank containing moderator – usually heavy water – and through which run pressure tubes.

CANDU: Canadian Deuterium Uranium reactor.

Cave: room with heavily shielded wall, within which highly radioactive materials can be handled by remote control.

CEA: Commissariat à l'Énergie Atomique (France).

CEGB: Central Electricity Generating Board.

Cerenkov radiation: blue light emitted when nuclear radiation travels through a transparent medium (like water) at a speed greater than that of light in the medium.

Chain reaction: see pp. 30–31.

China syndrome: possible consequence of core meltdown, when a molten mass of intensely radioactive material plummets through vessel and containment into the earth beneath in the direction of China (unless the reactor is in, say, Japan).

Cladding: sometimes just *Clad* (as noun): metal sheath (Magnox, zircaloy, stainless steel or ceramic) within which reactor fuel is hermetically sealed.

Containment: structure within a reactor building – or the building itself – which acts as a barrier to contain any radioactivity which may escape from the reactor itself.

Contamination: radioactivity where it should not be.

Control rod: rod of neutron-absorbing material inserted into reactor core to soak up neutrons and shut off or reduce rate of fission reaction.

Conversion ratio: number of fertile nuclei converted to fissile, compared to number of fissile nuclei lost by undergoing fission.

Coolant: liquid (water, molten metal) or gas (carbon dioxide, helium, air) pumped through reactor core to remove heat generated in the core.

Cooling pond: deep tank of water into which irradiated fuel is discharged upon removal from a reactor, there to remain until shipped for reprocessing or long-term storage.

Core: the region of a reactor containing fuel (and moderator, if any) within which the fission reaction is occurring.

Critical: refers to a chain reaction in which the total number of neutrons in one 'generation' of a chain is the same as the total number of neutrons in the next 'generation' of the chain; that is, a system in which the neutron density is neither increasing nor decreasing.

Criticality: the state of being critical.

Criticality accident: inadvertent accumulation of fissile material into a critical assembly, accompanied by outburst of neutrons and gamma radiation.

Cross-section: hypothetical 'target-area' measuring the probability of a nuclear event.

Crud: impurity deposit inside a reactor.

Curie: amount of radioactive material giving off 37 000 million radioactive emissions per second; radioactivity of 1 gram of radium.

Daughter, or *Daughter product*: the substance into which a radioactive nucleus transforms itself by radioactive decay.

Decay: radioactive transformation.

Decay heat: heat generated by radioactivity in the fuel of an operating reactor; additional to heat from chain reaction, and cannot be shut off.

Decontamination: transfer of unwanted radioactivity to a less undesirable location.

Densification: compaction of fuel inside cladding, as a consequence of irradiation; can lead to fuel damage because of unbalanced internal and external pressures.

Depleted uranium: uranium with less than the natural proportion (0·7 per cent) of uranium-235, which has been removed in an enrichment process and transferred to the remaining 'enriched' uranium.

Deuterium: hydrogen-2, heavy hydrogen; its nucleus consists of one proton plus one neutron, rather than the one proton only of ordinary hydrogen.

Deuterium oxide: heavy water: water in which the hydrogen atoms are heavy hydrogen.

Deuteron: nucleus of heavy hydrogen.

Disassembly: structural damage within a reactor core as a result of an excessive release of energy; 'blowing apart'.

Divergence: the achievement of criticality; 'going critical'.

Diversion: euphemism for theft, as applied to strategic or 'special' nuclear material.

DOE: Department of Energy (UK or US).

Dose: amount of energy delivered to a unit mass of a material by radiation travelling through it.

Dose-rate: time rate at which radiation delivers energy to unit mass of a material through which it is travelling.

Doubling time: time taken for a breeder reactor to produce additional fissile material enough to duplicate its total 'pipeline inventory' (see p. 71).

Drywell: on a boiling water reactor, the concrete containment around the reactor pressure vessel.

ECCS: emergency core-cooling systems.

Electron: negatively charged particle; much lighter than proton or neutron.

Enriched, as in *Enriched uranium*: uranium in which the proportion of uranium-235 has been increased above the natural 0·7 per cent.

Enrichment: process of making enriched uranium.

ERDA: Energy Research and Development Administration; one of the two US Federal agencies created after the split-up of the AEC; later became the Department of Energy (DOE).

Excited: having an excess of energy.

Exposure: to radiation: passage of radiation through a material.

Fall-out: radioactive fission products created by nuclear explosions, which descend from the atmosphere onto the surface of the earth.

Fast: of neutron: high energy, direct from fission.

Fast breeder: reactor designed to have conversion ratio greater than one, using unmoderated fast neutrons.

FBR: fast breeder reactor.

Fertile: of material like uranium-238 or thorium-232, which can by neutron absorption be transformed into fissile material.

Fissile: capable of undergoing fission.

Fission: rupture of a nucleus into two lighter fragments (*Fission products*) plus free neutrons – either spontaneously or as a consequence of absorption of a neutron.

Flux: of neutrons, moving cloud of particles, particularly in reactor core: number of neutrons through unit area in unit time.

FOE: Friends of the Earth.

Fuel: material (such as natural or enriched uranium or uranium and/or plutonium dioxide) containing fissile nuclei, fabricated into a suitable form for use in a reactor core.

Fuel pin: single tube of cladding filled with pellets of fuel.

Fuel rating: instantaneous power output per unit mass of fuel; measured as kilowatts per kilogram of uranium; also known as *Specific power*.

Fusion: the combination of two light nuclei to form a single heavier nucleus.

Gamma ray: high-energy electromagnetic radiation of great penetrating power emitted by nucleus.

Gas centrifuge: uranium enrichment device by which heavier uranium-238 nuclei are slightly separated from lighter uranium-235 nuclei by centrifuging of uranium hexafluoride gas; full-scale plant uses many thousands of centrifuges in cascade.

Gaseous diffusion: uranium enrichment process utilizing slight difference in rate of diffusion of uranium-235 and -238 hexafluoride molecules through porous metallic membrane; full-scale plant uses many thousands of diffusion cells in cascade.

Gigawatt: one thousand million watts.

Graphite: black compacted crystalline carbon, used as neutron moderator and reflector in reactor cores.

Gray (Gy): unit of radiation exposure.

G S: *Girdler-sulphide*: process used for the production of heavy water.

Half-life: period of time within which half the nuclei in a sample of radioactive material undergo decay; characteristic constant for each particular species of nucleus.

Heat exchanger: boiler, in which hot coolant from reactor core raises steam to drive turbogenerator; see also *Intermediate heat exchanger*.

Heavy hydrogen, Heavy water: see *Deuterium, Deuterium oxide*.

Helium: light chemically inert gas used as coolant in high temperature reactors.

Hex: uranium hexafluoride, easily vaporized uranium compound used in enrichment processes.

High-level: of radioactive waste, intensely radioactive with medium to long half-life.

H T G R: high temperature gas-cooled reactor.

I A E A: International Atomic Energy Agency.

ICGNE: International Consultative Group on Nuclear Energy.

ICRP: International Commission on Radiological Protection.

INFCE: International Nuclear Fuel Cycle Evaluation.

Intermediate heat exchanger: tube array in a sodium-cooled reactor in which hot radioactive primary sodium coolant transfers heat to non-radioactive secondary sodium coolant.

Iodine: as iodine-131; biologically hazardous fission product of short half-life (8 days) which tends to accumulate in the thyroid gland.

Ion: atom shorn of one or more electrons, and therefore electrically charged.

Ionizing radiation: radiation which can deliver energy in a form capable of knocking electrons off atoms, turning them into ions.

Irradiated: of reactor fuel, having been involved in a chain reaction, and having thereby accumulated fission products; in any application, exposed to radiation.

Isotope: form of an element, with the same number of protons in its nucleus as all other varieties of the element, but a different number of neutrons from the other varieties of the element.

JCAE: Joint Committee on Atomic Energy of the US Congress.

Kilowatt: one thousand watts.

Krypton: a chemically inert gas; the isotope krypton-85 is a troublesome fission product at present released to the atmosphere from reprocessing plants.

KWU: Kraftwerk Union (Federal Republic of Germany).

Laser enrichment: separation of uranium-235 from -238, selective excitation of one isotope with a laser; potentially a short-cut to highly-enriched uranium, which would present a serious problem as regards possible misuse of fissile material; applicable also to plutonium to separate 239 from higher isotopes.

Light water: ordinary water – to distinguish it from heavy water.

LIS: laser isotope separation – that is, *Laser enrichment*.

LMFBR: liquid metal fast breeder reactor.

Load following: varying the power level of a reactor to match requirements of an electricity distribution system.

LOCA: loss-of-coolant accident.

Low-level: of radioactive waste, not particularly radioactive.

LWR: light water reactor – either pressurized water reactor or boiling water reactor.

Magnox: alloy used as fuel cladding in first-generation British gas-cooled reactors, which are therefore called Magnox reactors.

Manhattan project: the 'Manhattan District' of the US Army Corps of Engineers – code name for the project which developed the atomic bomb.

McMahon act: The US Atomic Energy Act 1946, which banned any further transfer of nuclear information from the USA to the erstwhile allies Britain and Canada, and set up the US Atomic Energy Commission (AEC) and the Joint Congressional Committee on Atomic Energy (JCAE).

Megawatt: one million watts.

Meltdown: of reactor core, consequence of overheating which allows part or all of the solid fuel in a reactor to reach the temperature at which cladding and possibly fuel and support structure liquefy and collapse.

Mixed oxide: of reactor fuel, fuel in which the fissile nuclei are plutonium-239, mixed with natural or depleted uranium in a proportion equivalent to enriched uranium.

Moderator: material whose nuclei are predominantly of low atomic weight (e.g. light water, heavy water, graphite) used in reactor core to slow down fast neutrons to increase probability of their absorption in uranium-235 or plutonium-239 to cause fission.

MUF: material unaccounted for; refers to discrepancy between amount of fissile material expected at any point in the fuel cycle and amount actually measured; may indicate that *diversion* has occurred.

MWe: megawatts electric.

MWt: megawatts thermal.

NaK: sodium–potassium alloy with low melting point, used as coolant in early fast breeder reactors and as emergency coolant in some later designs.

Neutron: uncharged particle, constituent of nucleus – ejected at high energy during fission, capable of being absorbed in another nucleus and bringing about further fission or radioactive behaviour.

NNPA: Nuclear Non-Proliferation Act (US).

NPT: Non-Proliferation Treaty, intended to control the spread of nuclear weapons and their technology.

NRC: Nuclear Regulatory Commission, successor to the AEC with responsibility for licensing nuclear facilities in the USA.

NRDC: Natural Resources Defense Council (USA).

NRPB: National Radiological Protection Board (UK).

NSSS: nuclear steam supply system – in a nuclear power station, everything up to but not including the turbogenerators: reactor and its facilities (refuelling machine, control installation, fuel handling bay, cooling pond, steam generators if applicable etc.).

Nuclear reactor: see p. 31ff.

Nucleon: either proton or neutron.

Nucleus: see p. 24.

Nuclide: nucleus of isotope; nuclear species.

Off gas: radioactive gas from within a reactor which is released to the atmosphere, usually after a delay to reduce its radioactivity.

Period: of reactor, time taken for a certain increase (or decrease) of power level; short period makes a reactor difficult to control.

Pile: formerly nuclear reactor – after the first reactor, Chicago Pile No. 1.

Pipeline inventory: total amount of fissile material associated with one reactor: amount in operating core, in cooling pond, in reprocessing plant, in fuel fabrication plant and in transit.

Plowshare: US programme for civil engineering applications of nuclear explosives.

Plutonium: heavy artificial metal, made by neutron bombardment of uranium; fissile, highly reactive chemically, extremely toxic alpha-emitter.

Power density: in a reactor core, heat output per unit volume; measured in kilowatts per litre.

PRA: probabilistic risk assessment.

Pressure suppression pool: in a boiling water reactor, circular tunnel at bottom of *drywell*, half-filled with water, to condense steam from reactor cooling system if necessary.

Pressure vessel: large container of welded steel or pre-stressed concrete within which are reactor core and other reactor internals.

Pressurizer: in a pressurized water reactor, electrically heated boiler in a cooling system which boils water as necessary to maintain coolant pressure.

Price–Anderson Act: US Act of Congress limiting the third-party insurance liability of reactor operators in the event of an accident, and providing Federal indemnity to this limit.

Proton: positively charged particle, constituent of nucleus.

PTB: Partial Test Ban – treaty banning tests of nuclear weapons in the atmosphere.

Purex: plutonium–uranium extraction; original technology for reprocessing of irradiated reactor fuel.

PWR: pressurized water reactor.

QA: Quality Assurance.

Rad: radiation absorbed dose; measure of exposure to radiation.

Radiation, Nuclear: neutrons, alpha or beta particles or gamma rays which radiate out from radioactive substances.

Radioactivity: behaviour of substance in which nuclei are undergoing transformation and emitting radiation; note that *radioactivity* produces *radiation* – the two terms are *not* equivalent.

Radiogenic: caused by radiation, as certain types of disease.

Radioisotope: radioactive isotope.

Radionuclide: radioactive nuclide.

Radium: intensely radioactive alpha-emitting heavy element.

Radon: alpha-emitting radioactive gas given off by radium.

Reactivity: measure of ability of assembly of fissile materials to support sustained chain reaction. *Coefficient of Reactivity*, measure of the way the reactivity of an assembly changes in response to any other change, as for instance of temperature.

Reflector: of neutrons, a material of low atomic weight (light or heavy water, graphite) around a reactor core to reflect neutrons back into the reaction region.

Refuelling: replacement of reactor fuel after it has sustained maximum feasible *burn-up*; necessitated by loss of *reactivity*, build-up of neutron-absorbing fission products, and cumulative damage from radiation, temperature, coolant etc.

Rem: Roentgen equivalent man: unit of radiation exposure, compensated to allow for extra biological damage by alpha particles or fast neutrons.

Reprocessing: mechanical and chemical treatment of irradiated fuel to remove fission products and recover fissile material.

Runaway: accidentally uncontrolled chain reaction.

Running release: planned emission of radioactive material to the outside air or water.

Safeguards: term applied to keeping track of special nuclear material to prevent *diversion*.

Scram: emergency shutdown of fission reaction in reactor.

Separative work: measure of energy required to enrich uranium.

S G H W R: steam generating heavy water reactor.

Shielding: wall of material (concrete, lead, water) surrounding source of radiation, to reduce its intensity.

Sievert (Sv): unit of radiation exposure, compensated to allow for extra biological damage by alpha particles or fast neutrons.

S I P I: Scientists' Institute for Public Information (U S A).

S N M: *Special nuclear material*: fissile material potentially usable in nuclear weapons.

Specific activity: radioactivity per unit mass.

Specific power: heat output per unit mass of fuel; see *Fuel rating*.

Steam generator: boiler, in which hot coolant from reactor raises steam to drive turbogenerator.

Strontium: isotopes, particularly strontium-90; fission products, biologically hazardous beta-emitters.

Sub-critical: insufficiently supplied with neutrons to sustain a self-propagating chain reaction.

S W U: *Separative work* unit.

Tailings: fine grey sand, left over from extraction of uranium from ore; it contains radium, emits radon.

Tails assay: amount of fissile uranium-235 left in uranium depleted during enrichment process.

Thorium: fertile heavy metal.

T M I, T M I-2: Three Mile Island (pp. 169–72).

Tritium: hydrogen-3 – nucleus contains one proton plus two neutrons; radioactive.

Uranium: heaviest natural element, dark grey metal; isotopes 233 and 235 are fissile, 238 fertile; alpha-emitter.

Vitrification: fusing of high-level waste into glass-like solid.

WASH-740: AEC document *Theoretical Possibilities and Consequences of Major Accidents in Nuclear Power Plants* (1957).

WASH-1250: AEC document *The Safety of Nuclear Power Plants (Light Water Cooled) and Related Facilities* (1973).

WASH-1400: AEC document *An Assessment of Accident Risks in US Commercial Nuclear Power Plants* (1974).

Watt: measure of rate of transfer of energy; an adult human being gives off between 100 and 200 watts of heat.

Wigner energy: energy stored in graphite moderator as a result of deformation by radiation.

WISE: World Information Service on Energy.

Xenon poisoning: accumulation of neutron-hungry fission product xenon-135, reducing reactivity of reactor.

Yellowcake: mixed uranium oxides, with formula U_3O_8, produced from uranium ore by extraction process in uranium mill.

Zircaloy: alloy of zirconium used as fuel cladding; has low cross-section for absorption of neutrons.

Appendix B

Ionizing Radiation and Life

In Chapter 1 (pp. 25–8) we described how nuclear activities can produce four types of 'ionizing radiation': alpha, beta and gamma radiation and neutrons. There and elsewhere we indicated briefly some of the evidence accumulated since the discovery of radioactivity, about the effects of ionizing radiation on living organisms, including human beings. The study of these effects is called 'radiobiology'. It is a subject of deep and controversial complexity, the more so since humanity began to create radioactivity in quantity. It is far beyond the scope of this book to describe in detail the findings of radiobiology. However, since most of the immediate potential nuclear hazards arise from the effect of ionizing radiation on living things, some radiobiology is essential to pinpoint contentious issues.

In what follows 'radiation' means 'ionizing radiation' – not, for instance, sunlight (see p. 23).

It is agreed that, fundamentally, ionizing radiation is not good for you. The passage of alpha, beta or gamma radiation or neutrons through living tissue transfers energy to the atoms and molecules of the tissue, in a way which is bound to be more or less disruptive to the delicate organization of living systems. The disruptive effect is roughly proportional to the 'linear energy transfer' (LET) of the radiation. Beta and gamma radiation are of low linear energy transfer, alpha and neutron radiation of high linear transfer – also depending on the energy of the radiation. However, unlike a bullet in the brain, radiation – except in massive doses – is comparatively subtle in its effects. The radiation is invisible, and so, in almost all cases, is the damage. However, after radiation energy has disrupted some molecules in a living cell, the cell's biochemical behaviour may be affected. Gradually, instead of playing its accustomed role in metabolic activity, breaking down appropriate substances, building up others, the system goes awry. Some substances may no longer be broken down, but allowed to accumulate; others may be created in error, further disturbing the system's biochemistry.

Living systems have built-in redundancy, extra systems to take over when some fail; they can also carry out a considerable amount of repair work on deranged sub-systems. Under some circumstances radio-biological damage is taken care of without ever becoming evident. But in others the initial disruptive effect precipitates consequences that multiply the disruption. Unfortunately, we still do not know exactly how the initial radiation damage triggers successive harmful consequences in living tissue. It is clear that a massive dose of radiation can simply overwhelm a living system, with so much primary damage that it is incapable of recovery. But much more insidious harm can be done even by a single alpha particle, beta particle, gamma ray or neutron, although it may take a very long time to develop: years or even decades. Its final manifestation may be totally unrecognizable as radiation damage, not only because of the time-lag but also because the pathological outcome may result from a very long train of cumulative biological consequences triggered by a random jolt to a minute but sensitive component. It can be appreciated, then, that the scientific study of radiobiology is challenging, frustrating, and open to widely differing interpretations of data.

The body whose standing is highest in the field of radiobiology as it affects decision-making is the International Commission on Radiological Protection (ICRP), founded in 1928. It is made up of leading radiobiologists from many different countries; its committees meet regularly to assess the current understanding of radiobiological phenomena. On the basis of such assessments, the ICRP proposes standards for fields where radiation effects may arise. The ICRP's reports and recommendations form the basis of radiobiological standards in virtually all countries engaged in nuclear activities, although they are construed and applied differently in, for instance, the USA and the UK. ICRP Publication 26 is a concise presentation of the Commission's recommendations for radiation standards, and the reasoning upon which such standards are based. Other ICRP publications provide more detail, data and analysis.

Radiation effects can be described as acute or late, depending on whether they show up within a matter of weeks of the radiation exposure, or only perhaps years afterwards. The effects can be further classified as 'somatic' and 'genetic'. A 'somatic' radiation effect shows up in the organism – perhaps the human body – which has been exposed. A 'genetic' effect shows up in the offspring or later descendants. Acute effects are fairly easy to identify as radiation injuries; late somatic effects may be much harder, and genetic effects may not be identifiable at all.

Acute radiation injury – hundreds of rads in a short time – causes damage to the tissues which form red blood cells; very high doses may also damage the stomach and intestines, and extreme doses the central nervous system. But smaller doses usually entail a longer sequence of biological consequences. Leukaemia may be induced five years or more after the exposure; other cancers may not show up until as much as twenty years after the exposure. Cataracts may form on the eye; there may be skin damage. Fertility may be impaired. At the level of virtual undetectability is 'non-specific ageing' or 'radiation life-shortening', whose basis is very obscure. These are all somatic effects.

Even a single gamma ray can cause damage to a reproductive cell, to a gene or chromosome. If the damaged cell then participates in the formation of an offspring, the effect of the damage appears in the offspring – or, possibly, only in later generations. If the damage is sufficiently serious the offspring may not survive; if it does survive to reproduce, the so-called 'mutation' may slowly become a widespread feature of the descendent population.

We are continually subjected to ionizing radiation from natural sources: cosmic rays; uranium and thorium in the earth; and certain radioactive isotopes of substances in our bodies, particularly potassium-40. This 'background' radiation varies considerably from place to place on the earth, and according to height above sea-level. It is usually of the order of 100 millirem (0·1 rem) per year. Since this natural radiation is inescapable, radiobiologists assume that we have learned to live with it biologically. This does not mean that it is harmless, merely that whatever harm it does allows us to exist without discernible ill-effects. Accordingly, the natural background is taken as a baseline for the setting of standards governing man-made radiation. The fundamental ICRP recommendations set down 'maximum permissible doses' for those occupationally exposed to radiation, and 'dose limits' for members of the general public. Since radiation workers are aware of the possible hazards, and are expected to be supervised accordingly, the ICRP allows them more exposure; to allow a margin, the dose limits for the public are set ten times lower than the maximum permissible doses for occupational exposure. The maximum permissible dose for workers is set at 50 mSv (5 rem) per year and the dose limit for the public at 5 mSv (0·5 rem) per year. Exposure limits for particular parts of the body are designated by 'weighting factors' totalling 1; the sum of the 'weighted' dose equivalents must not exceed 50 mSv. Weighting factors are: gonads 0·25; breast 0·15; lung and red bone marrow 0·12 each; thyroid and bone surface 0·03 each; and the remainder 0·30.

In the UK, when a proposal to release radioactivity is made, a survey determines where this radioactivity will go. Different radioisotopes follow different paths. When radioactivity is released into a waterway, some of it is deposited on the bed of the waterway, some washed ashore, some taken up by plants or by animals, and so on. A radioisotope which is initially dilute in the original discharge may be concentrated by organisms which absorb it. All these possibilities must be identified and assessed. For the planned discharge, a 'critical group' of ultimate human consumers is distinguished, those whose consequent exposure is likely to be greatest, for radioactivity following the 'critical path'. If the critical group's exposure is to be kept below the ICRP Dose Limit, this implies a restriction on the original discharge, called a 'derived working limit'. The maximum permitted rate of radioactive discharge is set accordingly, and regular monitoring ensures that the limit remains within the ICRP criteria.

In the USA a different approach is taken. Standards are set on a nationwide basis. For any particular radioisotope there is laid down a maximum permissible concentration in air (MCP_a) or water (MCP_w). No effluent discharged from a nuclear facility must exceed this concentration at the site boundary. Clearly, it is easier for some facilities than for others to meet these criteria; some nuclear installations discharge effluents whose concentrations are a significant fraction of the maximum permissible, while others discharge effluents far below the permissible concentrations.

At present, medical applications of radiation account for much the largest part of public exposure to man-made radiation. Medical and dental X-rays, and various forms of radiotherapy – notably radiation treatment for cancer – are instances in which a clear-cut benefit to the exposed individual is weighed against the statistical, varying, but small possibility of radiation injury. It is less easy to make a risk–benefit comparison in respect of other forms of man-made radiation. Fall-out from nuclear weapons tests is a measurable contributor to present-day radiation exposure to the public, for benefits which must be regarded as debatable. This brings us squarely into the realm of controversy. What – if any – deleterious effects are caused by the as yet far lower level of radiation resulting from running releases from civil nuclear installations? Are the radiation standards adequate to protect public health? Are they adequate to protect the health of workers in nuclear facilities? And how are the decisions made, and the standards set and enforced?

The ICRP is essentially a self-appointed, self-perpetuating standing committee. Its membership is drawn largely from those scientists who gained their experience working in national nuclear programmes, and who therefore must be supposed to have a predisposition in favour of these nuclear programmes. In the US the National Academy of Sciences, the scientific body of highest standing, has had for a decade a committee on the Biological Effects of Ionizing Radiation (BEIR). The reports of the BEIR Committee might be expected to be the last word on the subject. But the third report (BEIR III) was withdrawn directly after publication in 1980, and eventually republished in a revised form, modifying its conclusions in a way which made it more acceptable to the US nuclear industry. Professor Edward Radford, Chairman of the BEIR Committee, thereupon wrote a devastating critique, both of the revised report and of the way it had been handled.

The controversy centred on the assessment of the risk of cancer from exposure to radiation. Actual human data on radiation-induced cancer is sparse; most of it comes from the victims of the Hiroshima and Nagasaki bombs. Radford and some of his colleagues felt that the data-base should include all recorded instances of tumours in the victims; but the revised BEIR III report considered only cancers which led to death and were so recorded. The other main point of contention was the form of the assumed relationship between radiation exposure and cancer risk. The original BEIR report of 1972 assumed that any exposure, however small, entailed some risk, and that the risk increased in direct proportion to the exposure: a so-called 'linear no-threshold' model. But after much disputation, the revised BEIR III report adopted a 'linear-quadratic' relationship, which assumed that the risk from a low exposure was less than directly proportional to the risk from a higher exposure. Since most people receive only low exposures, this approach – in the view of Radford and many others – might lead to a serious underestimate of the total cancer hazard from low-level radiation.

Matters were further complicated by the lack of understanding of the mechanism by which radiation induces cancer, and by a belated realization that the Hiroshima and Nagasaki exposure data had been wrongly interpreted. Studies of the cancer rate of workers at the Hanford nuclear reservation in Washington State, and at a naval dockyard where nuclear submarines were fitted, led to further controversy, and to accusations that official support for these studies was withdrawn when the studies began to lead to embarrassing conclusions about occupational health hazards of radiation exposure to the workers. Clearly the question of the effects of ionizing radiation is not only a scientific but also a political question, and one of acute sensitivity.

On a broader front, criteria for man-made radiation must look forward to the quantities of radioactivity which planned nuclear developments entail. Studies of the 'dose commitment' of radioactivity already released indicate that margins may become very much narrower before the end of this century. For the moment the most pressing requirement is better data, more careful compilation of records of exposure and medical histories of various kinds. Radiation experts point out repeatedly that it is not satisfactory to extrapolate uncritically from experiments on animals to forecast effects on human beings. But since few people would countenance planned experimentation on human subjects it ought to be a basic tenet of radiation medicine to collect and compile detailed records of effects where they occur or may occur. To take an obvious instance: the USA established in 1968 a 'Transuranium Registry' to keep track of employees who had encountered plutonium and other actinides in the course of their employment, and to follow their subsequent medical histories. In the UK such record-keeping was undertaken only in 1975. Until better evidence is available radiobiology will remain a hotbed of profoundly frustrating controversy.

Appendix C

Bibliography: A Nuclear Bookshelf

For those wishing to learn more about nuclear reactors and their world, the following sources are variously useful. Many of them will in turn indicate others: the nuclear bookshelf has expanded enormously in the last decade. See Appendix D for addresses of organizations named.

The International Atomic Energy Agency (IAEA) publishes a wide range of material. For hard data the shelves filled with the *Proceedings* of the four Geneva Conferences on the Peaceful Uses of Atomic Energy, and the 1977 Vienna Conference on Nuclear Power and its Fuel Cycle, are invaluable, albeit technical. The IAEA Directory is a multi-volume compilation of data on the world's reactors; the paperback *Power Reactors in Member States* is a handy compendium. The IAEA also publishes the proceedings of the many specialized conferences it organizes, and a free and useful quarterly *Bulletin*. IAEA publications are available from its Division of Publications.

The OECD Nuclear Energy Agency publishes conference proceedings and reports, and an annual *Activity Report*. The United Nations Scientific Committee on the Effects of Atomic Radiation publishes a major survey on *Ionizing Radiation*; and the International Commission on Radiological Protection (ICRP), whose *Publication 26* lays down the bases for most national restrictions on radioactivity and radiation, also publishes more detailed standard international references on radiobiology. The Uranium Institute publishes proceedings of its international symposia, which cover far more than the narrow question of uranium supply and demand.

The US Atomic Energy Commission (AEC) during its twenty-eight years published an enormous range of material, from the unintelligible to the trivial, with a vast amount of essential information in between. Its successors, the Nuclear Regulatory Commission (NRC), and the US Department of Energy (for a short time the Energy Research and Development Administration – ERDA), have since taken on the task. The US DOE and NRC publish weekly compilations of news releases, regulatory guides, *Power Reactor Events*, the bi-monthly journal *Nuclear Safety*, reports on important developments, environmental impact statements – the list could be expanded for pages. The NRC maintains a Public Document Room at 1717 H Street, N.W., Washington DC, at which all their public documents are available for inspection; photo-copying facilities are also provided. Inquirers can often obtain free a single copy of a new DOE or NRC publication; the news releases give details. All such documents are available for purchase from the National Technical Information Service of the US Department of Commerce, 5285 Port Royal Road, Springfield, Va 22151. The United Kingdom Atomic Energy Authority publishes, among many other items, a monthly bulletin, *Atom* – free and excellent. Atomic Energy of Canada Limited (AECL) publishes *Ascent*, a full-colour quarterly. Other national nuclear authorities should also be noted as sources for information.

Other official sources include the various US Congressional hearings, too numerous to list, and, in Britain, those of the Parliamentary Select Committee on Science and Technology and its successor, the Select Committee on Energy. Background papers for the US Congress – again too numerous to list – are also worthy of note; they are obtainable from the Superintendent of Documents, US Government Printing Office, Washington DC 20402. Ask whether the particular topic in which you are interested has been covered. The Library of Congress in Washington is of course another invaluable repository of information; the Congressional Research Service at the Library provides penetrating up-to-date commentary on many nuclear issues.

Industrial and commercial associations of nuclear interests, like the Atomic Industrial Forum in the USA and the British Nuclear Forum, offer a variety of information, documents and periodicals. On a more academic level organizations like the American Nuclear Society, the British Nuclear Energy Society, and the Canadian Nuclear Association publish conference proceedings and reports of value.

Books on nuclear matters abound, from the esoteric to the embarrassing. Some of the more notable – but by no means all – are mentioned below. In the most general terms, they are grouped according to their focus, although the categorization can only be arbitrary and partial. After some books which offer one or another general overview of the nuclear scene, there follow those which deal mainly with nuclear technology itself: safety; low-level radioactivity and radiation; nuclear economics; energy strategy; and nuclear weapons proliferation.

The most concentrated compilation of basic nuclear physics and engineering is undoubtedly the *Sourcebook on Atomic Energy* by Samuel Glasstone (Van Nostrand Reinhold, Third Edition 1968): a one-volume encyclopedia, clear without avoiding technicalities. It does, however, omit most of the more troubling aspects of the story. Much more technical is *Nuclear Reactor Engineering* (Van Nostrand Reinhold, Third Edition 1981) by Samuel Glasstone and Alexander Sesonske, a standard work on the subject, loaded with valuable information and data, but very much a specialized textbook. *Elements of Nuclear Power* by D. J. Bennet (Longman, Second Edition 1981) is an excellent British textbook.

The official history of US nuclear development began with *A General Account of the Development of Methods of Using Atomic Energy for Military Purposes under the Auspices of the United States Government 1940–1945*, by Professor H. D. Smyth, published in August 1945 at the behest of Major General Leslie Groves, Director of the Manhattan Project. For obvious reasons the report is now known simply as the Smyth Report. Later editions were published by the Princeton University Press. The Smyth Report is a fascinating account of the physics and – to some extent – the politics of the first nuclear programme, fascinating especially in the light of subsequent knowledge. The official histories of the US Atomic Energy Commission and its forerunners are *The New World 1939–1946* by R. G. Hewlett and O. E. Anderson (Pennsylvania State University Press, 1962), and *Atomic Shield, 1947–1952* by R. G. Hewlett and F. Duncan (Pennsylvania State University Press, 1969). The various US official histories tend to gloss over controversies and present strictly 'official' interpretations. Not so the UK official histories by Margaret Gowing. *Britain and Atomic Energy 1939–1945* (Macmillan, 1964) describes the first days; then follows *Independence and Deterrence* (Macmillan, 1974), which recounts British nuclear development from 1945 to 1952. This is by any criterion an extraordinary book, massively detailed, yet readable, even gripping. Volume I deals with *Policy Making*, Volume 2 with *Policy Execution*. Together they describe how the UK acquired nuclear weapons virtually in secret, and how scientists and engineers created a vast industry with almost no guidance or direction by politicians. Despite its size and cost, this book is essential reading for anyone anxious to unravel the way nuclear energy has developed.

One of the first – and still one of the best – general historical accounts of the early years of nuclear development is *Brighter than a Thousand Suns* by Robert Jungk (Penguin, 1970). Subtitled *A Personal History of the Atomic Scientists*, it is a vividly readable narrative of the men who made the A-bomb and the H-bomb, in particular Robert Oppenheimer and Edward Teller. Ralph Lapp's classic *The Voyage of the Lucky Dragon* (Penguin, 1958) tells the story of the luckless Japanese fishermen caught by the fall-out from the Castle Bravo H-bomb test. This was probably the first important popular book to challenge the AEC's policies and their execution.

A French quasi-official view of early history is offered by Bertrand Goldschmidt, a nuclear pioneer and long-time director of the French CEA, in *The Atomic Adventure* (Pergamon/Macmillan, 1964). Goldschmidt's distinctive slant is brought up to date in *The Atomic Complex: a Worldwide Political History of Nuclear Energy* (American Nuclear Society, 1982).

Nuclear Power: its Development in the United Kingdom, by R. F. Pocock (Unwin/Institution of Nuclear Engineers, 1977), is a readable and thoughtful account. *The Nuclear Power Decisions*, by Roger Williams (Croom Helm, 1980), is a well-mannered but uncompromising critique of Britain's nuclear cock-up; *Nuclear Power and the Energy Crisis: Politics and the Atomic Industry*, by Duncan Burn (Macmillan, 1978) is a ferocious polemic on the same subject – good on the details of the AGR fiasco, gullible on the putative advantages of the PWR.

The Great American Bomb Machine, by Roger Rapoport (Dutton, 1971), is a brilliant dissection of the most unpleasant innards of the AEC's nuclear weapons activities. *The Atomic Establishment*, by H. Peter Metzger (Simon & Schuster, 1972), is a scathing historical critique of the AEC in action, and of its cosy relationship with the Congressional Joint Committee on Atomic Energy, the watchdog that 'did nothing in the night-time'. *The Nuclear-Power Rebellion: Citizens vs. the Atomic Industrial Establishment*, by Richard Lewis (Viking, 1972), is an inside look at the many battlefronts upon which the US civil nuclear industry and its critics have clashed. The author, as Editor of the *Bulletin of the Atomic Scientists*, had a unique opportunity to observe the confrontations and indeed to further them. *Citizen Groups and the Nuclear Power Controversy*, by Stephen Ebbin and Raphael Kasper (MIT Press, 1974), analyses how public participation in US nuclear licensing procedures has worked – or failed to work – in general, taking a closer look at three hearings including the one on emergency core-cooling systems.

Much the best of the critical histories recently published is *The Nuclear Barons*, by Peter Pringle and James Spigelman (Michael Joseph, 1982). As the title indicates, the authors are concerned about the power and influence of the world's nuclear elite. They tell their story in fast-moving, immediate prose; and they make a good, if occasionally heavy-handed, case for concern, backing it up with a vast panoply of notes and references. Also highly recommended is *The Politics of Uranium* by Norman Moss (André Deutsch, 1981). Moss is an American journalist based in Britain, long concerned with nuclear affairs; he brings to the subject a perceptive eye and a distinctive, sceptical viewpoint, without grinding any axes.

Two complementary and authoritative perspectives on the nuclear scene are provided by the sixth report from the Royal Commission on Environmental Pollution, *Nuclear Power and the Environment* – popularly known as the Flowers Report; and the Ranger Uranium Environmental Enquiry First Report – popularly known as the Fox Report. These two reports appeared only a month apart, in autumn 1976; each offers a thorough and dispassionate survey of civil nuclear energy in its most controversial aspects. The reports are complementary in that the Flowers Report addresses itself in particular to the problems created by the production and possible use of plutonium, while the Fox Report considers the related problems of uranium. They should be side by side on any serious nuclear bookshelf; and they should be well thumbed.

Nuclear Power for Beginners, by Stephen Croall and Kaianders Sempler (Writers and Readers/Beginners Books, 1978), has no such respectable pedigree; indeed its polemical cartoon-poster format makes it look downright disreputable. But it is withal a lively, technically accurate and witty – if also scurrilous – onslaught on nuclear activities: highly entertaining but with obvious serious undertones.

Undoubtedly the most authoritative work on reactor problems is *The Technology of Nuclear Reactor Safety* (MIT Press, 1965 and 1973), edited by Theos Thompson and J. G. Beckerley of the AEC. It covers *Reactor Physics and Control* in Volume 1, and *Nuclear Materials and Engineering* in Volume 2. Although frighteningly expensive they are the most comprehensive, indeed exhaustive compilation available, with a wealth of historical information both technically detailed and scrupulously complete. Thompson and Beckerley pulled no punches; when poor design, short-sightedness, corner-cutting or carelessness led to trouble they were uncompromising. The first popular book on reactor problems was probably Sheldon Novick's *The Careless Atom* (Delta, 1970). It is still one of the best – readable, purposeful and businesslike.

By the mid 1970s the question of nuclear safety had become folklore; so much so that the *Reader's Digest* commissioned a book about it: *We Almost Lost Detroit*, by John G. Fuller. Fuller took as his theme the sorry life-history of the Enrico Fermi-I fast breeder power plant, but augmented it with additional material from around the world: a breezy, readable, popular guided tour of the underside of the nuclear story, which upset the industry badly, not least because of its antecedents.

Petr Beckmann, in turn, upset nuclear critics with a rumbustious knock-about polemic entitled *The Health Hazards of Not Going Nuclear* (Golem Press, 1976). When he could not find a publisher for it, he published it himself – and forthwith sold many thousands of copies, especially to the nuclear industry itself, which distributed the book widely as a counter to the wilder excesses of 'anti-nuclear' literature.

The accident at Three Mile Island released a torrent of paper. Official commentaries included the report from the President's Commission (the Kemeny Report), the report from the NRC Special Enquiry (the Rogovin Report), and the report from the US Senate. Of these the pick is probably the first volume of the Rogovin Report, which narrates the story of the accident minute by minute, in what would be a hilarious black comedy, if it were not both true and frightening. Unofficial Three Mile Island books include *Three Mile Island: Prologue or Epilogue* by Daniel Martin (Ballinger, 1980), *Three Mile Island: the Hour by Hour Account of What Really Happened* by Mark Stephens (Random House, 1980) and *Three Mile Island: Thirty Minutes to Melt-Down*, by Daniel Ford (Penguin, 1981). Of these the pick is probably the last, placing the accident in the context of the US nuclear safety scene overall; Ford, one-time executive director of the Union of Concerned Scientists, has been a leading critic for more than a decade.

While the world was preoccupied with events in Pennsylvania, Zhores Medvedev, an eminent Russian geneticist living in exile in London, published *Nuclear Disaster in the Urals* (W. W. Norton, 1979). It is a remarkable scientific detective story, piecing together many strands of evidence to prove that there was a catastrophic release of radioactivity in the Soviet Union in 1957 or early 1958, presumably at the secret nuclear weapons complex near Chelyabinsk in the Urals.

Medvedev's story has attracted bitter challenge from Western nuclear establishments, but an unbiased reader will find his arguments all but irrefutable. Sooner or later we shall know the truth, and Medvedev's efforts will have led the way.

Sir Alan Cottrell, former Chief Scientific Adviser to the UK government, does not like to be regarded as a nuclear critic. In *How Safe is Nuclear Energy?* (Heinemann, 1981) he makes clear his enthusiasm for nuclear energy and his impatience with those less enthusiastic. However, given that he is a metallurgist of international standing, his doubts about the safety of PWR pressure vessels – once again noted here – are nuclear criticism of a high order, and have been so treated.

The effects of low-level radiation are discussed in the publications of the ICRP and the BEIR reports (see pp. 232–35), and in the publications of the UK National Radiological Protection Board. *Biological Effects of Radiation* by J. G. Coggle (Wykeham, 1973) is an excellent primer on the subject. *Nuclear Power, Man and the Environment* by R. J. Pentreath (Taylor & Francis, 1980) is a more recent book which addresses also the policy issues arising, without being anodyne. John Gofman, a former senior AEC scientist, has since become one of the fiercest critics of nuclear activities, with several books to his name, most recently *Radiation and Human Life*, a 900-page *vade mecum*. Ernest Sternglass takes a yet more radical view in *Secret Fall-Out: Low-Level Radiation from Hiroshima to Three Mile Island* (McGraw-Hill, 1981); his analysis is acutely alarming, but it is challenged by many radiobiologists not otherwise very sympathetic to the nuclear establishment.

Since the mid 1970s economic issues have come to the forefront of nuclear controversy. *Light Water: How the Nuclear Dream Dissolved* by Irvin Bupp and Jean-Claude Derian (Basic Books, 1978) is a dispassionate account of how the US established its domestic nuclear industry by first sowing the nuclear seed in Europe; the authors declare themselves firmly in favour of nuclear energy, but deeply unhappy about the way it has been managed. Charles Komanoff has published a series of penetrating analyses of the economics of nuclear plants, most recently *Power Plant Cost Escalation: Nuclear and Coal Capital Costs, Regulation and Economics* (Komanoff Energy Associates, 1981). His analyses demonstrate that even when subject to very stringent environmental constraints, coal-fired plants have been and will continue to be more economic than nuclear plants; the findings are, needless to say, hotly disputed by the nuclear industry, but amply documented and carefully argued.

David Lilienthal was the first chairman of the US AEC, but became in later years one of its most trenchant critics. *Atomic Energy: a New Start* (Harper & Row, 1980) was his last book, a lucid and fervent plea for the industry to put its house in order. When the industry lost Lilienthal, they lost a statesman they badly needed.

It is impossible to list even a fraction of the material on 'energy strategy' which has appeared within the last decade. Official and quasi-official reports in English include *World Energy: Looking Ahead to 2020*, from the World Energy Conference (IPC, 1978); *Energy in Transition: 1985–2010*, from the US National Research Council Committee on Nuclear and Alternative Energy Systems (CONAES) (W. H. Freeman, 1979); and *Energy in a Finite World*, from the International Institute for Applied Systems Analysis (Pergamon, 1981). All these and many other international and governmental reports foresee a major role – some indeed a crucial role – for nuclear electricity in the future. The Swedish Secretariat for Futures Studies published *Energy in Transition* (1976) and *Solar Versus Nuclear* (Pergamon, 1980), thoughtful and crucial analyses of the social impact of energy choices and constraints: highly recommended, as is *Energy Choices in a Democratic Society* (US National Academy of Sciences, 1980), which was omitted from the final version of the CONAES report. A trail-breaking challenge to the official view was *World Energy Strategies* by Amory Lovins (Ballinger, 1975); it was followed by *Soft Energy Paths* (Penguin, 1977): intense, concentrated, and by no means easy reading, but a key source of innovative energy thinking. Its effects can be seen, for instance, in *Energy Future*, edited by Robert Stobaugh and Daniel Yergin (Random House, 1979), a report from the Energy Project at the Harvard Business School. Similar studies in other industrial countries suggest a similarly limited future role for nuclear energy. The official studies listed above also proclaim the importance of nuclear energy for the Third World; but *Future Energy Consumption of the Third World* by Markus Fritz (Pergamon, 1981) suggests otherwise. The author contacted the responsible organizations in 156 countries to ascertain their intentions, and found very few that assigned any priority whatever to nuclear development: a remarkable study, revealing and readable.

Since its foundation in 1966 the Stockholm International Peace Research Institute has been the most authoritative international source of information on nuclear weapons; it has also returned repeatedly to the problem of proliferation. Its annual Yearbook, *World Armaments and Disarmament* (Taylor & Francis), offers regular if depressing up-dates on the situation; and its one-off publications include several important studies, notably *Nuclear Energy and Nuclear Weapons Pro-liferation* (Taylor & Francis, 1979). The basic problem was definitively described in *Nuclear Theft: Risks and Safeguards* by Mason Willrich and Theodore Taylor (Ballinger, 1974), a classic study of an eviscerating subject. John McPhee wrote a profile of Taylor for the *New Yorker*; the resulting book, *The Curve of Binding Energy* (Dutton, 1974), is a translation of *Nuclear Theft* into human terms, grimly convincing and offering little compensatory consolation. In McPhee's view the next nuclear explosion in a city seems to be only a matter of time – and not a very long time. *The Last Chance: Nuclear Proliferation and Arms Control* by William Epstein (Free Press, 1976) is an urgent personal history of the efforts to get a grip on the problem, written by a vastly experienced and deeply committed Canadian diplomat. The Con-gressional Research Service of the US Library of Congress has pre-pared a series of *Nuclear Proliferation Factbooks* which summarize key data and reproduce significant documents. *Nuclear Power Issues and Choices* (Ballinger, 1977), a report from a group of the senior US experts, provided the arguments underlying the attempt by the Carter administration to defer the commercial use of plutonium. Called the 'Ford–MITRE' report in reference to its institutional sponsors, its analysis has become even more apposite in the intervening years. *The Plumbat Affair* by Elaine Davenport, Paul Eddy and Peter Gilman (André Deutsch, 1978), recounts the extraordinary story of one par-ticular evasion of international safeguards, in which Israel acquired 200 tonnes of uranium yellowcake while eluding both the IAEA and Euratom; the episode was thereafter hushed up for a decade before it was at last uncovered. *Energy and War* by Amory and Hunter Lovins (Friends of the Earth, 1980) is probably the most uncompromising analysis of the issue to date; its authors conclude that only the complete progressive elimination of civil nuclear activities overall can minimize the grey area from which the gravest threat of proliferation arises.

In a field changing as swiftly as that of nuclear energy, periodical publications also constitute essential sources of information. The monthly *Bulletin of the Atomic Scientists* (5801 S. Kenwood Avenue, Chicago, Ill. 60637) was founded in 1945; throughout nearly four decades it has been a forum for consistently far-sighted and thoughtful discussion of the implications of nuclear energy for the world. The weekly *New Scientist* (Commonwealth House, 1–19 New Oxford Street, London WC1A 1NG) is a lively and readable magazine covering all aspects of science and society, with regular dispatches about energy technology and policy, including nuclear energy. *Nature* (4 Little Essex Street, London EC1) and *Science* (1515 Massachusetts Avenue, N.W., Washington DC 20005) weekly, include, as well as original scientific papers and notes, regular news items which include major coverage of the nuclear field. The quarterly journal *Energy Policy* (PO Box 63, Westbury House, Bury Street, Guildford, Surrey GU2 5BH), although expensive, contains valuable in-depth articles by leading authorities on energy in general and nuclear energy in particular. Then there are the three periodicals which should be known to all those concerned with nuclear energy. *Nucleonics Week* is an international weekly newsletter published by McGraw-Hill (1221 Avenue of the Americas, New York, NY 10021); devoid of advertising and often sufficiently outspoken to bring aggrieved looks from the industry, but its subscription price is more than $ 600 per year ... check your library. *Nuclear News*, published monthly by the American Nuclear Society, covers the whole field of nuclear energy in depth and – considering its origins – gives balanced commentary on issues and controversies. *Nuclear Engineering International* is a glossy monthly, offering issue-length features on new nuclear plants and on particular aspects of the technology; its annual output includes an issue devoted to an exhaustive international list of firms offering nuclear material and services and an issue with a detailed directory of the world's reactors. It also publishes fold-out wall-charts of nuclear plants and other information.

Nuclear opponents, too, have their periodicals. The US Friends of the Earth publish the monthly *Not Man Apart* which includes the Nuclear Blowdown, a survey of important nuclear news items and other developments. The Nader organization publishes a monthly tabloid, *Critical Mass*. In the UK the Scottish Campaign to Resist the Atomic Menace (SCRAM, of course) publishes a monthly *Energy Bulletin*, and so on. The World Information Service on Energy (WISE) publishes a monthly *Bulletin*, and newsletters in many countries belonging to the net.

Appendix D

Nuclear Organizations Pro and Con

International Atomic Energy Agency, PO Box 100, A-1400 Vienna, Austria.

OECD Nuclear Energy Agency, 38 Boulevard Suchet, F-75016 Paris, France.

United States Nuclear Regulatory Commission (Washington DC 20555) and US Department of Energy (Washington DC 20585) now embody the former US Atomic Energy Commission.

United Kingdom Atomic Energy Authority, 11 Charles II Street, London SW1, UK.

Commissariat à l'Énergie Atomique, 31–33 Rue de la Fédération, F-75752 Paris, France.

Atomic Energy of Canada Ltd, 275 Slater Street, Ottawa K1A OS4, Ontario, Canada.

(For other national nuclear organizations, government and industry, try embassies for details.)

Uranium Institute, New Zealand House, Haymarket, London SW1, UK.

United Nations Scientific Committee on the Effects of Atomic Radiation, United Nations, New York, USA.

International Commission on Radiological Protection, Clifton Avenue, Sutton, Surrey, UK.

National Radiological Protection Board, Harwell, Didcot, Oxfordshire OX11 0RQ, UK.

Atomic Industrial Forum, 7101 Wisconsin Avenue, Washington DC 20814, USA.

British Nuclear Forum, 1 St Albans Street, London SW1Y 4SL, UK.

American Nuclear Society, 555 North Kensington Avenue, La Grange Park, Ill. 60525, USA.

British Nuclear Energy Society, 1–7 Great George Street, London SW1P 3AA, UK.

Canadian Nuclear Association, 111 Elizabeth Street, Toronto, Ontario M5G 1P7, Canada.

Stockholm International Peace Research Institute, Sveavaegen 166, S-113 46 Stockholm, Sweden.

Pugwash, 9 Great Russell Mansions, 60 Great Russell Street, London WC1, UK.

Friends of the Earth Ltd, 377 City Road, London EC1V 1NA, UK.

Friends of the Earth Inc., 1045 Sansome Street, San Francisco, California 94111, USA.

Friends of the Earth International, Box 7235, S-40235 Goteborg, Sweden.

Anti-Nuclear Campaign, PO Box 216, Sheffield S1 1BD, UK.

SCRAM, 30 Frederick Street, Edinburgh EH2 2JR, Scotland.

Greenpeace, Damrak 83, 1012 LN Amsterdam, The Netherlands.

World Information Service on Energy, Czaar Peterstraat, 1018 N.W. Amsterdam, The Netherlands.

Scientists' Institute for Public Information, 30 East 68th Street, New York, NY 10021, USA.

Natural Resources Defense Council, 1725 First Street, N.W., Washington DC 20006, USA.

Union of Concerned Scientists, 1384 Massachusetts Avenue, Cambridge, Mass. 02238, USA.

Critical Mass Energy Project, PO Box 1538, Washington DC 20013, USA.

Nuclear Information and Resource Service, 1536 Sixteenth Street, N.W., Washington DC 20036, USA.

Index

See also Appendix A: Nuclear Jargon, Appendix C: Bibliography, and Appendix D: Nuclear Organizations. Names – such as US, UK, USSR, Canada, uranium, plutonium, AEC, AEA, PWR and BWR – which occur repeatedly throughout the text are not listed separately but only under other specific references.